CHEMICAL ANALYSIS
OF
FISH POND SOIL AND WATER

Chemical Analysis of Fish Pond Soil and Water

G.N. Chattopadhyay

2026
Daya Publishing House®
A Division of
Astral International Pvt. Ltd.
New Delhi - 110 002

First Published, 1998
Reprinted, 2026

ISBN: 978-81-7035-177-1 (Hardbound)

Publisher's note:

Every possible effort has been made to ensure that the information contained in this book is accurate at the time of going to press, and the publisher and author cannot accept responsibility for any errors or omissions, however caused. No responsibility for loss or damage occasioned to any person acting, or refraining from action, as a result of the material in this publication can be accepted by the editor, the publisher or the author. The Publisher is not associated with any product or vendor mentioned in the book. The contents of this work are intended to further general scientific research, understanding and discussion only. Readers should consult with a specialist where appropriate.

Every effort has been made to trace the owners of copyright material used in this book, if any. The author and the publisher will be grateful for any omission brought to their notice for acknowledgement in the future editions of the book.

Published by : **Daya Publishing House®**
A Division of
Astral International Pvt. Ltd.
– ISO 9001:2015 Certified Company –
4736/23, Ansari Road, Darya Ganj
New Delhi-110 002
Ph. 011-43549197, 23278134
E-mail: info@astralint.com
Website: www.astralint.com

Digitally Printed at : Replika Press Pvt. Ltd.

**Dedicated to
the memory of
my parents**

FOREWORD

Fish constitutes a major share of dietary proteins to millions of Indian people. It has been estimated that by the end of the present century India will need about 13 million tons of fish of which about 7 million tons will come from culture fisheries. With the potential area for fish culture being rapidly used up for other purposes, there is an urgent need to develop technologies so that the productivity of our culture system can be substantially increased.

The productivity of any fish pond depends largely on the quality of bottom mud which is the store house of nutrients. Various chemical and biochemical changes continuously take place in the underwater pond mud resulting in release of nutrients to the overlying water. The growth of fish food organisms and thus finally the growth and yield of the end product, namely, fish depends greatly on the presence of these essential nutrients in water in adequate and balanced quantities.

To increase the productivity, manures and chemical fertilisers are some times added to the ponds. Their efficiency also depends to a great extent on the quality of pond soil.

In the above contexts, the importance of chemical analysis of fish pond soil and water can be hardly over-emphasized. The present manual for fish pond soil and water analysis will thus go a long way to help in understanding the fish pond productivity and developing judicious fertilisation programme for increasing the same. I am sure the book will serve very useful purpose to all those concerned with various aquaculture systems.

Kalyani
West Bengal

Professor L.N.Mandal
Former Vice-Chancellor,
Bidhan Chandra Krishi Viswavidyalaya

PREFACE

Increasing attentions are now being paid towards soil and water quality management of different fish culture systems, especially in fish ponds, with a view to improving their productivity potentials. This is usually achieved through monitoring of soil and water properties of the ecosystem at various stages. Adequate knowledge about different analytical procedures for assessment of these properties are the basic pre-requisites for these exercises. There are numbers of good publications on chemical analysis of fish pond soil and water and most of them deal with the subject in very comprehensive manner. Although such literatures are highly useful for the researchers and other persons involved in studies on chemistry of the aquatic ecosystems, yet much of the knowledge are usually not so essential for most of the workers engaged in fisheries development without much background in chemistry. During my long association with fisheries research, I had been feeling the need of a hand book on analyses of fish pond soil and water which will be useful to those people, having only moderate knowledge in chemistry, to carry out their day to day analytical work with good accuracy and considerable ease. In this book, therefore, I have tried to describe various procedures of different chemical analyses in such a manner so that they become easily accessible to any one who is interested in the subject and is having only some basic knowledge in chemistry.

I shall be happy if this hand book can prove itself useful to the aquaculturists during the courses of their exercises with analysis of fish pond soil and water. Any suggestion or comment for improving the usability of the book will be highly appreciated.

My family members had been constant source of encouragement during the preparation of the manuscript. Staff members and research scholars of the soil testing laboratory of Institute of Agriculture, Visva-Bharati also extended all kinds of help to this effort.

I express my sincere thanks to all. I take this opportunity to express my gratitude to my teacher Professor L.N. Mandal, an eminent soil scientist and former Vice-Chancellor, Bidhan Chandra Krishi Viswavidyalaya for his kindness of writing the Foreword for the book.

G.N. Chattopadhyay

CONTENTS

1

INTRODUCTION

Estimation of different nutrient elements and environmental parameters, which are associated with productivity of various aquaculture systems, forms an integral part of efficient management of any fish pond. In this book attempts have been made to describe the methods of analysis of some of these parameters magnitude and dynamics of which are required to be understood for adopting day to day pond productivity management practices and also to find out solutions in case of emergence of problems. The major objective of this book is to provide necessary technology to the fisheries scientists and technicians to take up exercises on soil and water monitoring without going into details of chemical components of such programmes. The methods described in this publication have, therefore, been carefully chosen to be simple, easy to perform and, at the same time, efficient ones which may be adopted without much background in chemistry. However, for a better understanding of the subject matter, a brief discussion about the general properties of pond soil and water has been attempted in this pre-fatorial section. The idea of introducing this chapter is to provide some brief information about the chemical parameters dealt in this book, which will have little significance to the reader unless correlated with biological properties.

1.1 Importance of Soil and Water in Aquaculture

It is well known that productivity of a fish pond depends largely on the abundance of fish food organisms and also on occurrence of congenial environmental condition in the pond. The first step in the food chain (Fig. 1) of a fish pond is constituted by

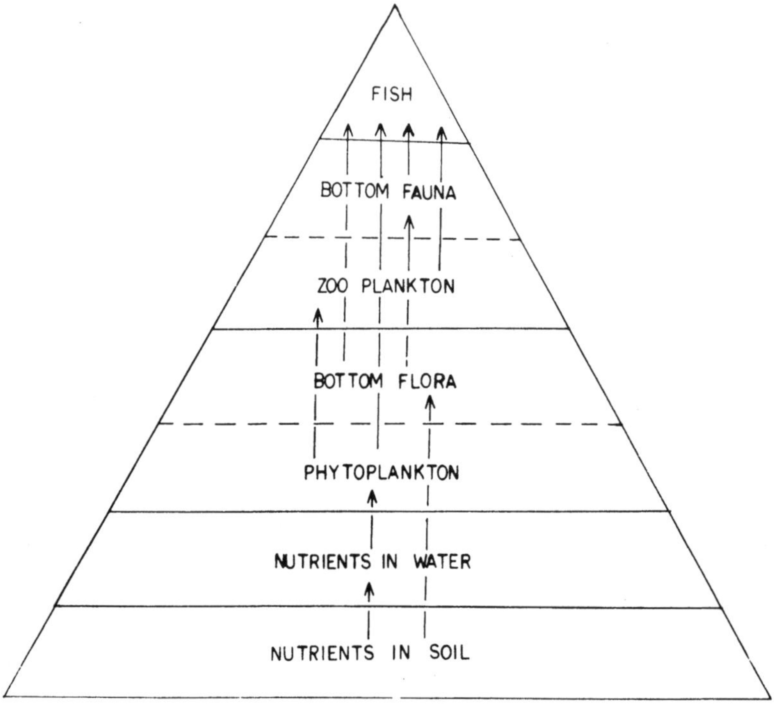

Fig. 1: Food Chain in Fish Ponds

primary food organisms e.g. phytoplankters, which derive their nutrients from the pond environment and with the help of solar radiation undergo photosynthetic activities. Occurrence of these nutrients in pond water and maintenance of its relevant chemical condition depend largely on the nature and properties of the bottom soil wherein a series of chemical and biochemical reactions continuously take place resulting in release of different nutrients in overlying water and also their absorption in the soil mass (Mandal and Chttopadhyay, 1992). Considering this importance of bottom soils in maintaining productivity of fish ponds Hickling (1971) described such soils as the "Chemical laboratory of the fish pond".

In view of this importance of overlying water as well as bottom soil in determining the productivity of a fish pond, an intimate knowledge about the nature and properties of these two phases need to be understood thoroughly for developing a clear idea about the ecosystem and obtaining, therefrom, good production of fishes.

1.2 Characteristics of pond soil and water

The single major phenomenon which differentiates a fish pond soil from an upland one is the presence of a layer of standing water above the bottom soil. As a result of this, oxygen supply to the bottom soil becomes restricted and a reduced or anaerobic condition develops below a thin oxidised layer in the bottom soil surface (Pearsall and Mortimer, 1939). Characteristics of these two layers are entirely different. The oxidised or aerobic layer in the soil-water interphase is characterised by presence of oxidised forms of various nutrient elements viz., nitrate (NO_3^-), sulphate ($SO_4^=$), ferric iron (Fe^{+++}) etc and also aerobic group of micro-organisms. The lower reduced or anaerobic layer, on the other hand, exhibits reduced forms of the elements viz. ammonium (NH_4^+), sulphide ($S^=$), ferrous iron (Fe^{++}) etc. and also anaerobic group of microorganisms. The magnitude and dynamics of these oxidation-reduction reactions play an important role in governing most of the chemical, biochemical and microbial behaviours in the pond soil and water and also in maintaining congenial environmental conditions (Mandal and Chattopadhyay, 1990). Important behaviours of some of these parameters influencing productivity of fish culture systems have been discussed here.

1.2-1 pH and relevant properties

The major changes associated with electro-chemical properties of fish pond soil and water are reflected in pH and also the electrical conductivity (EC) values. The pH of acid soils usually tends to increase and that of an alkaline soil tends to decline with increase in the period and extent of submergence (Ghildyal, 1976). In addition, pH values of water exhibit marked diurnal variation especially in case of productive ponds owing to release of oxygen and consumption of carbon dioxide during photosynthetic activity of primary food organisms in day time and steady increase in concentration of carbon dioxide during night hours. Importance of pH of fish pond soil and water on pond productivity has been emphasized by many workers (Alikunhi, 1957; Hickling, 1971; Banerjea, 1967 and others) and this parameter is generally considered to be the most emphasized one for maintaining pond productivity. Neutral to slightly alkaline pH ranges for both soil and water are considered to be congenial for aquatic production owing to greater availability of most of the nutrient elements and also due to increased biological activities under this pH range (Boyd, 1978).

Under acidic conditions of fish ponds, liming is widely practised to raise the pH of pond soil and water. On the other hand, under extremely alkaline situation, gypsum may be required to reduce the pH to a congenial level. However, application of lime or gypsum above or lower the required amount may be either ineffective or even harmful to productivity of ponds. In addition to studies on pH of pond soil and water, therefore, methodologies for determining lime requirement (LR) and gypsum requirement (GR) appear to be important.

1.2-2 Soluble salts and associated parameters

Submergence tends to increase the solubility of many elements in the soil phase and thus helps to increase electrical conductivity (EC) of pond soil and water. Under freshwater conditions these changes do not usually reach very high values so as to influence the productivity through effects of total soluble salts. However, this behaviour may encourage more solubility of some particular ions in some cases and thus may alter ionic equilibrium of the system (Mandal and Chattopadhyay, 1990).

Like EC salinity also refers to amount of soluble salts in water and is of special significance in case of brackishwater aquaculture where highly saline tidal water is used for culture of fishes and prawns and variation in salt content influences the productivity of the culture system considerably. The total concentration of all dissolved substances in water is referred as total dissolved solids (TDS). Generally, concentrations of salinity and TDS are similar (Boyd 1978). All these three properties are interrelated and hence one parameter can be calculated from the value of the other with fairly good level of accuracy.

1.2-3 Primary nutrient elements

Among different nutrient elements, nitrogen (N), phosphorus (P) and potassium (K) are considered to be primary nutrients owing to their major roles in plant nutrition (Dev, 1995). Of them, P is generally considered to be the most critical single factor in the maintenance of pond fertility (Jhingran, 1975) owing to its utmost importance to primary fish food organisms and also occurrence of very low amount of this element in available form. This problem of availability of P in fish pond soil and water is due to fixation of phosphate ions into insoluble compounds of calcium (Ca) under alkaline environments and iron (Fe) and aluminium (Al) under acidic conditions resulting in release of very low amount of P in available form under either situation (Banerjea and Mandal, 1965). In view of this importance of transformation of P into different insoluble forms in fish pond soil, not only the methods of estimation of water soluble and available soil P are important but also the methods of determination of different inorganically bound forms viz. saloid bound, Al-bound, Fe-bound and Ca-bound forms need to be known. Although total P does not influence fertility status of fish ponds, yet the method of estimation of total P has also been described in this book in case the reader wants to assess the P supplying potential of the pond soil.

Nitrogen (N), the second primary nutrient element in fish pond nutrition, becomes available to primary fish food organisms in two inorganic forms viz. ammonium (NH_4^+) and nitrate (NO_3^-). N in the bottom soil remains mostly in organic form and, therefrom, is gradually mineralised to inorganic (NH_4^+) and (NO_3^-) forms in which N becomes available to primary producers in fish pond soil and water (Mandal and Chattopadhyay, 1992). In this book, methods of estimation of easily mineralisable organic nitrogen in the soil has been described as a measure of available N supplying capacity of the pond soil. Apart from this, determination of readily available inorganic forms of N viz. NH_4^+ and NO_3^- forms in pond soil and also in water have been described. Nitrite (NO_2^-), an intermediate form in N transformation, usually occurs in very small amount. However, under restricted availability of oxygen in pond bottom, concentration of this form of N may increase and become toxic to aquatic lives. Determination of NO_2^- form of N in pond water has, therefore, been included in the text for monitoring water qualities under polluted conditions.

Inspite of forming the third primary nutrient element influencing pond productivity, little work have so far been done to understand the chemistry of potassium (K) in fish ponds. High clay and organic matter content of productive fish pond soils (Singh and

Ram, 1971), associated with alkaline pH values usually tend to maintain moderate to good amount of K in pond soil and water (Chattopadhyay, 1995). This has probably restricted detailed studies on this nutrient element and the subject has been dealt by only a few workers (Golterman, 1967; Dobbins and Boyd, 1976 and others). In fish ponds, K becomes available to fish food organisms in watersoluble form in water and in watersoluble as well as exchangeable form in soil. Determinations of all these forms of K have been included in this book.

1.2-4 Presence of base elements

In addition to N, P and K, three other elements viz. calcium (Ca), magnesium (Mg) and sulphur (S) constitute the forum of major nutrient elements to be released primarily by the soils and are called as secondary nutrients. Apart from providing nutrition to fish food organisms, the first two elements along with sodium (Na) and also K serve as important components of base saturation of the pond soil and contribute to maintenance of alkaline pH in the pond environment. Like K, these base elements become available in watersoluble and exchangeable forms.

Total concentrations of all divalent cations is total hardness of water (Boyd, 1982) and is expressed in terms of ppm of calcium carbonate ($CaCO_3$) equivalent. Since Ca and Mg are the two major divalent cations in most of the fish ponds, total hardness also provides an assessment of occurrence of these two base elements in water.

1.2-5 Major anions in fish ponds

Chloride (Cl^-), sulphate ($SO_4^=$), bicarbonate (HCO_3^-) and carbonate ($CO_3^=$) constitute the major anions in fish pond ecosystem. In addition, hydroxyl (OH^-) ion occasionally registers its presence in appreciable amount, especially under highly alkaline conditions. $CO_3^=$ and HCO_3^- ions provide CO_2 in water which is required for photosynthesis. These two ions and also OH^- form the major sources of alkalinity and can be estimated through determination of this parameter. In the soil phase occurrence of $CO_3^=$ and HCO_3^- in the form of calcite, dolomite, etc can be assessed from the amount of alkaline earth carbonate which is also expressed as free calcium carbonate. In addition, watersoluble forms of these anions and also Cl^- and $SO_4^=$ ions can be determined from soil-water extract. Apart from being an important nutrient element, Cl^- ion remains associated with majority of soluble salts in soil and water and hence estimation of this ion helps to assess the total quantum of soluble salts to a fairly accurate degree. Sulphate is a major source of sulphur (S), an important constituent of protein. However, under highly reduced condition $SO_4^=$ may be converted to sulphide ($S^=$)form and affect productivity of the ecosystem (Mandal and Chattopadhyay, 1990). Although not included under regular analyses, determination of $S^=$form of sulphur deserves attention under some specific conditions.

1.2-6 Organic compounds in pond soil and water

A wide range of organic substances occur in pond water. Such materials include compounds like sugars, starches, amino acids, peptides, proteins, tanins, humic acids etc concentrations and relative distribution of which depend on several factors. These dissolved organic matters release considerable amount of nutrient elements to soluble form and thus aid to fertility status of ponds. High concentrations of organic matter, on the other hand, may consume large amount of oxygen from the pond water during its course of mineralisation and thus create oxygen tension in the aquatic environment. If this demand for oxygen becomes very high, anaerobic condition may develop in the pond, especially in the bottom, resulting in development of unhygeinic situation. In view of these effects of organic compounds on pond environment, the methods of determination of dissolved organic matter content of pond water appears to be important. In addition, determination of chemical oxygen demand (COD) of water, which indicates total amount of oxygen required for oxidation of all the organic matter in a water sample, provides a measurement of organic matter in water. The rate of oxygen consumption by plankton and bacteria in the water sample within a specific time and condition is expressed as biochemical oxygen demand (BOD) which also generates an idea of organic load in water. Phenol is a compound which often occurs in considerable amount in organically rich water and requires to be estimated under various environmental impact assessment programmes. Although not confined to organically polluted waters always, concentrations of free chlorine (Cl_2) sometimes need to be tested as a component of water pollution. Methods of determination of all these parameters have been dealt in this book.

Fish pond soils, in general, exhibit comparatively higher amount of organic matter (OM) than corresponding upland soils (Singh and Ram, 1971). This OM plays an important role in determining productivity of fish ponds. Apart from influencing various physico-chemical properties of the bottom soils alongwith releasing different nutrient elements to available form in pond environment, it also controls the oxidation-reduction reactions in the ecosystem (Mandal and Chattopadhyay, 1992). However, all the components of organic matter in soil are not equally active and, thereby, useful for productivity of soils (Jackson, 1971). In this book, therefore, determination of easily oxidisable organic matter of soil (Walkley and Black, 1934) has been described.

1.2-7 Some relevant physical parameters

Although this book intends to emphasize on analysis of various chemical properties of fish pond soil and water yet apart from such analyses only, a few physical parameters, which tend to influence productivity of ponds to a considerable extent, have also been included in the text.

When solar energy enters into the epilimnion and passes through it in the form of light, it heats up the surrounding water body and increases its temperature. Magnitude of this increase depends on concentration of various dissolved and suspended materials in water. Since most of the chemical and biological activities in a pond environment are accelerated or retarded by variations in temperature, this parameter plays an important role in determining productivity of ponds. Turbidity is a property which depicts the visibility of water. High turbidity of water, created by suspended particles, interferes with penetration of sunlight in pond water and, thereby, affects photosynthetic activity of primary fish food organisms. On the other hand, Boyd (1978) described turbidity caused by planktonic biomass to be desirable to aquaculture, to a certain extent.

Texture of a soil indicates relative occurrence of sand, silt and clay particles in it. Since finer particles are more active than the coarse ones, the magnitude of activity of the soil phase is determined largely by its textural composition. In addition, this property also controls, to a considerable extent, the water retentivity of the soils which is an important factor for undertaking aquaculture in any pond. Bulk density (BD) and particle density (PD) are two important parameters of soils to indicate their density and also to determine occurrence of pores in the soils. In view of the importance of all these physical properties in aquatic productivity, determinations of such properties have been included in the text.

1.2-8 Biological productivity of fish ponds

Apart from observations on fish growth, productivity potential of a fish pond can also be assessed from the population of phytoplankters which remain in the bottom of the food pyramid and, thereby, control the abundance of most of the natural fish food organisms in water. Concentrations of these primary fish food organisms can be estimated either by directly counting their population or by measuring their weight per unit volume of water and also chemically by estimating the rate of carbon assimilation during their photosynthetic activity i.e. primary productivity (Odum, 1971) or from the amount of chlorophyll concentration in water. Close correlation between chlorophyll-a value of water and abundance of phytoplankton has been reported by Dust and Shindala (1970) and many others. Both of these chemical estimations of pond productivity have been described in this book.

1.3. Epilogue

It will be observed from the above discussion that all attempts have been made in this publication to include the methodologies for determination of most of the chemical parameters of pond soil and water related with productivity of fish ponds. During his period of research in the field of fish pond soil and water chemistry, the author has observed the problems faced by large sections of fisheries personnel in undertaking chemical analyses of such soil and water. Inadequate aquaintance with the subject and also non-availability of simple forms of methodology often restrict their scope of research as well as day to day work related with assessment of pond soil and water qualities. This problem can only be encountered by providing them easier forms of methodologies without going into details of their chemistry part. While writing this book, therefore, keen attention has been paid to this aspect. At the time of describing various methodologies, care has been exercised to describe them in a manner which may be easily grasped by any person with only moderate background in chemistry. Chemical formulae for different chemicals had to be mentioned but they have been preceded, atleast once, by the full name for the chemical so that the user does not have to search for a dictionary of chemistry. In each section, requirements of instruments, chemicals and glasswares etc have been enlisted before going into methodologies. Procedural part has also been described step by step without going into details of chemistry of it. A brief outline of the principle involved in the analysis programme has been given at the beginning of each section for meeting the demand of interested readers. Implications of the observed results have been discussed at the end of most of the sections for enabling the worker to interprete the observed results through information on critical and optimum levels of different parameters. An appendix part has also been included at the end of the book to provide some useful information related to preparation of reagents, analysis of samples, calculation of data and interpretation of results. It is hoped that the book will be found useful by the readers.

2

WATER ANALYSIS

In fish culture, water quality is usually defined as the suitability of water for survival and growth of fishes and it is normally governed by only a few variables (Boyd, 1978). Most of these properties can be measured with appreciable degree of accuracy and the data can be used for the purpose of productivity management of the pond. Evaluation of water quality is carried out under a few stages viz. collection of water samples, preservation, analysis and interpretation. None of these components of water quality assessment is less important than the others and adequate cares should be exercised during the entire stretch of the work programme.

2.1 Collection of water samples

It is to be categorically borne in mind that collection of representative water sample is an integral part of the total analytical procedure and it should be given much more importance than is usually done. Since only a very small amount of water sample is collected from the entire pond area for subsequent analysis, every effort should be taken so that it becomes a true representative of the water body under investigation. Any negligence or irregularity during the period of sampling may ultimately lead to untrue representation of the water quality. No amount of caution or accuracy employed during the period of analysis of the water sample will be able to repair the damage to any useful extent. Thus the entire exercise may become futile due to slight carelessness during sampling period. So, for getting a correct picture of water quality of a fish pond, it must be ensured that the collected water sample reflects the quality of the water body.

2.1–1 Procedures of water sampling

For collection of water sample, the location of sampling should be determined first. Maximum photosynthetic activity of primary fish food organisms are observed in the upper layers of pond water and hence this zone should be given major attention. At the same time, some emphasis should be given to factors like direction of wind, availability of sunlight, hour of sampling etc. Wind action may change plankton population and composition from one location to another and thus may influence some of the chemical properties of water at different zones. Also, for the same reasons, shady zones of the water area should be separated out. Hour of sampling is another important criteria to be carefully considered. Due to effect of sunlight on photosynthetic activity in a fish pond, dissolved oxygen (DO) levels as well as pH values of water usually tend to increase with day time whereas the amount of free carbon di-oxide (CO_2) decreases. These behaviours may influence some other properties of pond water also. Considering these phenomena, collections of water samples are usually done during early morning hours when DO and free CO_2 values are likely to remain in critical concentrations. If, due to any reason, it is not possible to collect water samples at dawn, the hour of collection should be kept fixed for all regular sampling activities because deviation from routine hour of collection may not yield comparable results.

For collecting water samples, good quality glass or polythene containers of various sizes are used. For the purpose of DO estimation, bottles of around 125 ml volume may be taken while larger volumes of water in one or more containers may be collected separately for other analyses.

Sampling of surface water can be done by simply dipping the container slowly in water without disturbing the surface and avoiding development of any air bubble in the near vicinity of collection spot as well as in the container. For many specific purposes, it often becomes necessary to know the water quality at some particular depths. This can be achieved by using different water samplers generally available in the market. One such commonly used water sampler is "Kemmerer's Sampler", various modifications of which have been and may be made for different purposes. A schematic diagram of the sampler is presented in Fig. 1. This type of water sampler consists of a container (a) which can be closed with the help of a messenger (b) after filling in water from the desired depth. After collection, the sampler is pulled up and the collected water sample is transferred to the collection bottle through a draining pipe or tube (c) fitted with a clamp (d) after filling in water from the desired depth. After collection, the bottle is pulled up alongwith the sampler, removed from the structure and the collected water sample is used for analysis.

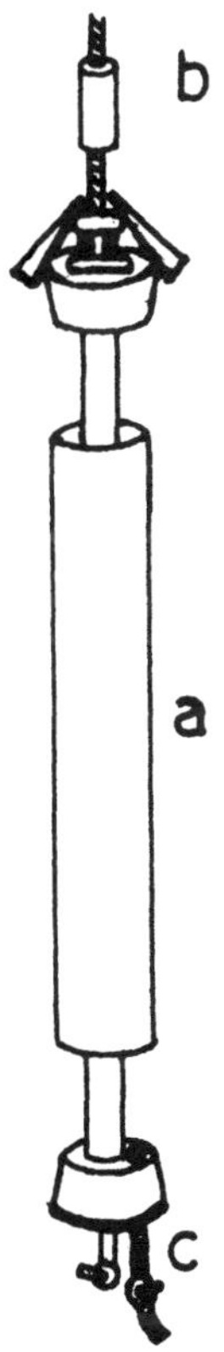

Fig. 2: Structure of a water sampler

2.1–2 Preservation of water samples

Some qualities of pond water may change quite rapidly on storing and hence adequate care should be taken to complete the water analysis as soon as possible after collection. Boyd (1978) described that a number of changes may occur among the constituents of water in sampling bottles including the absorption of substances onto the sides of the bottles, pH alteration through metabolic activities of micro organisms and the release of metabolites by micro organisms. This may lead to changes in DO and CO_2 concentrations, variations in alkalinity, transformation of different forms of nitrogen (N), viz. ammonification, nitrification, denitrification etc., transformation of phosphorus (P) and many other properties of water. However, it is not always possible to complete the entire analytical processes within a short period, especially when water samples are collected from places away from laboratory. Different preservatives are available for inhibiting changes in water qualities for a considerable period which provides some more time to the analyst for completing his analyses. Some guidelines for using preservatives for different analytical procedures have been presented in Table 1. This table will be useful to the reader for selecting suitable methods for preservation of water samples depending on his requirement.

Table 1 : Use of preservatives in chemical analyses of water

Parameter	*Preservative*	*Permissible time*
Acidity and Alkalinity	Cooling at 4°C	24 hours
Biochemical Oxygen Demand	—do—	6 hours
Carbon Di-Oxide	—do—	2 hours
Total Phosphorus	—do—	7 days
Chlorophyll–a	—do—	12 hours
Chemical Oxygen Demand	1ml Conc. $H_2SO_4\ell^{-1}$	7 days
Hardness	1 ml Conc. $HNO_3\ell^{-1}$	7 days
Ammonia and Nitrate	1 ml Conc. $H_2SO_4\ell^{-1}$	7 days
Soluble Phosphorus	Storage in I_2 treated bottles	7 days
Dissolved Oxygen (Winkler's method)	Fixing with Manganous Sulphate and Alkaline Iodide immediately	6 hours

2.2 Analysis of water sample

In this section, analytical procedures of some widely used parameters of water quality measurement have been discussed. Numbers of methods are available for estimation of some of these properties. However, considering that most of the laboratories engaged in fisheries research in different corners of the country are not usually very well equipped, attempts have been made to discuss the more common as well as simple methods for these determinations.

2.2.-1 pH

Importance of pH in determining pond productivity has been emphasized by many workers (Hickling, 1971; Jhingran, 1975 and others). The changes in pH in pond soil and water has practical significance in the maintenance of proper nutritional condition of the pond, apart from affecting the physiological functions of the fishes. It greatly affects the equilibria of hydroxide, carbonate, phosphate and silicate between bottom mud and overlying water and, thereby the precipitation of soluble nutrients and dissolutions of solid materials, sorption and dissorption of ions and the concentrations of aluminium, manganese, iron etc which have nutritional significance in fish ponds.

pH of water may be measured either colorimetrically or potentiometrically. The colorimetric methods involve less expenditure but may also be less accurate and need more time for determination than the electrical method (Golterman *et. al.* 1978). However, for many practical purposes, especially for on field studies, workers may have to employ colorimetric method. Both of these methods are being discussed here.

A. Colorimetric method

This method does not require any power supply and hence is preferred by many aquaculturists for estimating pH in the field itself, especially when the laboratory is not nearby and highly accurate results are not required.

Principle : The principle of colorimetric method of pH measurement is to develop pH dependant colour on the water sample by using pH indicator of specific range and to compare the developed colour with standard colour discs, colour charts or coloured buffers.

Requisites : (a) Apparatus : Lovibond comparator,

(b) Indicators : The indicators commonly used in such estimation with specific ranges of pH are as follows :

Indicator	*pH ranges*
Bromo phenol blue	3.0 — 4.6
Bromo cresol green	3.8 — 5.4
Bromo cresol purple	5.2 — 6.8
Bromo thymol blue	6.0 — 7.6
Phenol red	6.8 — 8.4
Cresol red	7.2 — 8.8
Thymol blue	8.0 — 9.6

Procedure

Generally Lovibond comparator is used for colorimetric estimations of pH values of water. Before using the comparator, a primary idea about the pH of the sample is usually obtained by dipping a strip of a wide range pH paper the change in colour of which provides an approximate idea about the pH of the sample. Then depending on the observed pH, select the pH indicator from the list of indicators and corresponding pH range. Attach the coloured disc of this indicator with comparator. Take 10 ml of clear water sample in two glass couvettes. Add 0.5 ml of the selected indicator to water in one couvette and keep it in the specific groove of the comparator. Put the other couvette, having water sample only, in another groove, just in front of the coloured disc. The disc has different shades of colours in it which are comparable to colours developed with the indicator at various pH values. Match the colour developed by the indicator in the water sample with the shades of colours of the disc by rotating it slowly. When the colour of the test sample and that of the disc matches, the corresponding reading gives the pH of water.

Comments

Colorimetric method of pH determination is difficult to be used for turbid water samples. Such water may often be made clearer by mixing a little amount of neutral barium chloride ($BaCl_2$), allowing it to settle and then decanting the clear liquid (Jhingran *et. al*, 1988).

Estimation with this method may be erroneous if concentrations of indicator in water sample do not remain uniform. Moreover, for determination of many water samples having wide ranges of pH values, frequent changes of the colour discs may be necessary. However, in spite of these problems, this method has remained popular for years especially in the remote areas without good facility of various equipments.

B. Potentiometric method

This method is more accurate than the colorimetric one, requires lesser time for determination and hence is widely recommended. Apart from electrically operated pH meters, portable battery operated pH meters are also now being used, especially in fields.

Principle

In potentiometric determination of pH, electromotive force of a galvanic cell is related to the activity of hydrogen ion in water. The voltage produced by the cell is converted by the potentiometer to a current which is proportional to the pH of the sample. The current is transmitted in terms of pH reading.

Requisites

(a) Apparatus — pH meter

(b) Glasswares — 100 ml beakers, 100 ml volumetric flasks

(c) Reagents — Buffer solutions :

Standard buffer tablets for different pH levels are available in the market. Three buffer solutions having pH 4.0, 7.0 and 9.2 are commonly used for calibrating pH meter. Usually one buffer tablet dissolved in 100 ml distilled water (DW) produces corresponding pH level of the solution. However, manufacturer's instructions are to be followed for preparing such buffer solutions.

Procedure

The operational procedure of the pH meter should follow the guidelines of the manufacturer. However, some basic principles of operations are as follows.

Keep the pH meter in "switch on" position for some time before use to allow a warming up time for the instrument. Then take about 50 ml of pH 7.0 buffer solution in a clean 100 ml beaker, deep the electrodes (electrode, if combined glass and calomel electrode is used) and calibrate the pH reading to 7.0. Remove the beaker, rinse the electrode (s) with DW, soak any adhering water in a filter paper and then dip into another buffer solution of either acidic or alkaline range, depending on the expected pH value of the water sample to be measured. Check if the second solution is giving the respective reading of the buffer. If any fluctuation is observed, calibrate the pH reading with the second buffer and again observe the pH reading in the first buffer solution by following the same method. In case water samples of wide range of pH values are to be analysed, the instrument may be calibrated with both ends of pH values (4.0 and 9.2).

When all the buffer solutions are giving exact readings of respective pH values, dip the electrode (s) into about 50 ml of water sample in a clean beaker and find the pH reading directly. In case of more than one sample, rinse the electrode(s) every time with DW, soak the moisture with a filter paper and go for next analysis. If large numbers of water samples are to be analysed, some intermediate checkings of the calibration are necessary. However, now a days, most of the instruments have a monitoring switch which indicates whether the calibration has shifted or not. After completion of the work, wash the electrode(s) with DW and keep dipped in a beaker containing DW.

Comments

Voltage fluctuation in power supply is a major problem generally observed in our country. This may affect the pH readings considerably. To overcome this problem, a sensitive voltage stabilizer may be attached to the instrument. For battery operated pH meters it should be ensured that power supplying capacity of the battery is intact. Another problem generally encountered in pH estimation is development of sluggishness of the electrode after prolonged use. In such case, the electrode should be dipped in 0.1 N hydrochloric acid (HCI) overnight, thoroughly washed and then used for further work.

C. pH values of water

Importance of pH of pond water in production of fishes has been studied for long in India, Alikunhi (1957) stated that waters on acid soils are usually less productive than those on alkaline soils. Banerjea (1967), while correlating productivity of large numbers of ponds with various soil and water properties, observed near neutral to slightly alkaline pH range to be most favourable for fish ponds.

2.2-2 Electrical Conductivity (EC)

Electrical conductivity (EC) value indicates the total concentration of ionised constituents of a water sample. It is closely related to the amount of total dissolved solids and is also used as an index of salt content of water. EC is reciprocal of electrical resistance and can be derived from the relationship $C = {}^{I}/_{R}$ where I is the current, R is the resistance and C is the conductivity. Jackson (1971) described specific conductivity of a solution as the conductance that would be measured at 25°C between electrodes 1 cm^2 in cross section and placed 1 cm apart and may be visualised as the conductance across a cm^3 or mmhos cm^{-1}. Although in recent years the units of conductance has been modified to dsm^{-1}, yet in the present book the more commonly used unit (mmhos cm^{-1}) has been used.

Principle

EC is determined with the help of an EC bridge, also known as conductivity meter. This instrument has a conductivity cell. The cell constant of this cell is determined first by measuring the conductance of standard potassium chloride (KCl) solution, specific conductance of which solution is known.

Cell constant = known sp conductance of standard KCl solution/conductance shown by the cell.

The conductance of the water sample is measured with this standardised instrument and the obtained conductance value multiplied by cell constant gives the EC of the sample.

Requisites

(a) Apparatus - Conductivity Meter.

(b) Glass Wares : Vol. Flasks 1 ℓ beakers 100 ml etc.

(c) Reagents :

(i) Standard (0.02 M) Potassium Chloride (KCl) solution : Dissolve 1.4912 g KCl in 1 ℓ of DW. EC of this solution is 2.39 mmhos cm^{-1} at 18°C and 2.768 mmhos cm^{-1} at 25°C.

Procedure

(a) Immerse the cell of the conductivity meter in the standard KCl solution and read the conductivity value in the meter.

(b) Calculate the cell constant from the relationship shown earlier using known specific conductance of the standard solution.

(c) Rinse the cell with DW and then immerse in the test water sample, read the conductivity as before.

Calculations

Calculate the EC of the water sample from the relationship- EC (mmhos cm^{-1}) = Measured conductivity (mmhos cm^{-1}) × cell constant.

Comments

EC values are closely related with temperature. EC of a solution increases approximately at the rate of 2 per cent per °C. Hence EC is generally expressed at a standard temperature of 25°C. Most of the EC meters are, however, having temperature compensation arrangement now a days so that direct EC reading for 25°C can be obtained.

Freshwater fish ponds, in general, exhibit low EC values which may be expressed as μ mhos cm^{-1} instead of mmhos cm^{-1} (mmhos cm^{-1} × 1000 =μ mmhos cm^{-1}) Boyd (1978) stated natural waters usually have EC values of 20 to 1500 μ mhos cm^{-1}. Some methods of calculating salt content of water from EC values have been presented in the section 2.2.3

2.2-3 Salinity

In fresh water fish culture, salinity of water usually remains very low and does not, in general, merit estimation. Exceptions are some ponds, especially those situated in the coastal areas, where water salinity some times creates problem and should be monitored regularly. This parameter, however, forms an essential component of water analysis in case of brackishwater aquaculture.

A. Estimation of total salt concentration

While working with brackishwater, it often becomes necessary to estimate the salinity of water in terms of total concentration of soluble salts. This is usually expressed as g of salts per ℓ of water or ppt (part per thousand). Salinity such expressed can be calculated fairly accurately from chlorinity or electrical conductivity value of water.

Determination of salinity from chlorinity

Estimation of Cl^- (ppm) has been discussed in section 2.2-10. Any value in ppm when divided by 1000, expresses the said value in ppt. e.g. if the observed chlorinity of a water sample is x ppm, the concentration, when expressed as ppt, will be x / 1000. For obtaining salinity value of water in ppt, the chlorinity should be determined by following the method described in section 2.2-10, the value be expressed as ppt and the following relationship be used.

Salinity (ppt) = Chlorinity (ppt) × 1.805 + 0.03

Determination of salinity from Electrical Conductivity

The salinity (ppt) of water can also be calculated approximately from the EC values of water by following the relationship – Salinity = Electrical Conductivity (mmhos cm^{-1}) × 0.64.

Estimations of equivalents of salts

The effects of salts on plants is more closely related to the equivalents of salts in the solution than gravimetric weight (Jackson, 1971). Hence in some cases it may become essential to estimate the equivalents of salts in unit volume of water. This can be achieved approximately from EC value by using the following relationship–

Meq of salt ℓ^{-1} = 12.5 × EC (mmhos cm^{-1}).

Comments

Salt content of freshwater ponds usually varies between 0.5 and 1.0 ppt. However, in arid areas or coastal zones the values may be higher. Most of the freshwater fishes usually grow well under water salinities up to 3.0 ppt while Ghosh et. al. (1973) observed 10.0 — 12.5 ppt to be the critical level for survival of Indian major carps.

2.2-4 Total Solids (TDS, TSS, TVS, TVDS)

Total solids of a water sample represent dissolved organic matters, particulate organic matters, dissolved inorganic substances excepting gases and suspended inorganic substances. Different forms of solids present in water e.g. total solids (TS), total dissolved solids (TDS), total suspended solids (TSS), total volatile solids (TVS) or total volatile dissolved solids (TVDS) can be measured by following the major principle of evaporation of the water sample.

Principle

TS of a water sample can be determined simply by evaporating a known amount of water sample and weighing the residue while TDS can be measured by the same manner but by using filtered water sample only. On the other hand, TSS is the difference between TS and TDS. TVS can be determined from the weight loss after ignition of the residue of TS analysis while TVDS can be determined by igniting the residue for TDS estimation.

For getting very accurate results for TDS or TVDS glass fibre filters are recommended. However, in absence of such facility, use of Whatman no. 42 filter paper will yield fairly good results. The second provision has been described in this chapter.

Requisites

(i) General : Porcelin basin, Whatman no. 42 filter paper, funnel, conical flask etc.

(ii) Muffle furnace

Procedure

For determination of TS, heat a clean porcelin basin in an oven at 105°C. Cool in a dessicator and weigh it accurately (W_1). Shake the water sample thoroughly and take 100 ml of the sample in the basin. Evaporate the water sample to dryness in oven at temperature around 100°C. After drying, keep the basin at 105°C for half an hour. Cool it in dessicator and weigh the basin alongwith the residue (W_2).

For estimation of TDS, filter the water sample with the help of a Whatman no. 42 filter paper and use 100 ml of the filtrate for drying and weighing (W_3) as before.

TVS can be determined by igniting the residue for TS estimation in a muffle furnace at 550°C for half an hour. Cool the basin in a dessicator and weigh it accurately (W_4).

Similarly for estimation of TVDS, ignite the residue in the basin for TDS analysis as above and get the weight of the ignited residue (W_5).

Calculation

$$\text{TS (ppm)} = (W_2 - W_1)\ \text{mg} \times 10$$

$$\text{TDS (ppm)} = (W_3 - W_1)\ \text{mg} \times 10$$

$$\text{TVS (ppm)} = (W_2 - W_4)\ \text{mg} \times 10$$

$$\text{TVDS (ppm)} = (W_3 - W_5)\ \text{mg} \times 10$$

Comments

An approximation of TDS value of water can be made by multiplying the respective EC value with a factor. Boyd (1978) suggested that this factor may be ranging between 0.55 and 0.90 depending on the composition of the dissolved solids.

2.2-5 Calcium and Magnesium

Calcium (Ca) and Magnesium (Mg) are two common cations which contribute to productivity of fish ponds not only as nutrient elements but also through their influence on hardness as well as alkalinity in water. One common and reliable method for determination of Ca and Mg in water is by complexometric titration using ethylene diamine tetra acetic acid (EDTA).

Principle

EDTA is a very efficient complexone which forms chelates with a number of polyvalent cations viz. Ca, Mg, F, Mn, Zn etc. This behaviour is utilised for determination of Ca by complexometric titration with EDTA using ammonium purpurate (Murexide) as indicator. Similarly, total amount of Ca and Mg can be determined through titration with EDTA after adding Eriochrome black T as indicator. Concentration of Mg in the water samples may then be determined from the difference of values of Ca + Mg and Ca alone.

Requisites :

Glass wares, 1 ℓ vol. flasks, burette, pipette, 100 ml conical flask etc.

Reagents :

(i) .01 N calcium solution : Dissolve 0.5005 g dried pure calcium carbonate ($CaCO_3$) in about 100 ml of approximately 0.2N HCI (Dilute 5.0 ml concentrated HCI to 250 ml with DW). Boil the solution for some time, cool it down and make up the vol. to 1 ℓ.

(ii) Ammonium purpurate (Murexide) indicator : Mix thoroughly 0.2 g murexide powder with 40 g potassium sulphate and store in a dark coloured bottle.

(iii) Eriochrome black T indicator : Dissolve 0.5 g eriochrome black T and 4.5 g hydroxylamine hydrochloride indicator in 100 ml ethyl alcohol. Add. 5 ml of 2 per cent sodium cyanide solution to it.

(iv) 4 N Sodium hydroxide (NaOH) : Dissolve 160 g NaOH in 1 ℓ DW. This solution need not be standardised.

(v) Ammonium chloride - ammonium hydroxide (NH_4Cl - NH_4OH) buffer solution : Mix 67.5 g NH_4Cl and 5.70 ml NH_4OH and dilute to a total volume of 1 ℓ.

(vi) 0.01 N EDTA solution : Weigh 2.0 g disodium dihydrogen ethylenediamine tetra acetic acid (Di-Na salt of EDTA). Dissolve in 1 ℓ of DW. Standardise the solution against standard 0.01 N Ca solution by following the method described below.

The buffer, EDTA and eriochrome black T solutions should be prepared fresh during use for a few days.

Procedure

Standardisation of EDTA solution : Take 5 ml 0.01 N Ca solution in a 100 ml conical flask, dilute to about 25 ml with DW. Add 5 ml 4 N NaOH solution and about 25 mg of murexide indicator. Mix the contents well to get a orange red colouration. Titrate it with 0.01 N EDTA solution against a white background untill the colour changes to purple.

The colour change is not very distinct and hence the titration during end point should be carried out cautiously. Find out the strength (factor) of EDTA solution by using the following relationship :

Factor = $^5/_V$ where V = Volume of EDTA solution required.

Determination of Ca

Take 5 ml of water sample in a 100 ml conical flask and proceed as in case of standardisation process.

Determination of Ca + Mg

Take 5 ml of water sample in a 100 ml conical flask and dilute it to about 25 ml as before. Add 1 ml of NH_4 Cl-NH_4 OH buffer and 3-4 drops of eriochrome black - T indicator to get a wine red colour. Titrate against 0.01 N EDTA untill the colour changes to blue.

Calculations

(a) Concentration of Ca (ppm) = ml of EDTA used × F × 40

(b) Concentration of Mg (ppm) = (ml of EDTA used for Ca + Mg - ml of EDTA used for Ca) × F × 24

where, F = factor of EDTA solution.

Comments

In calcium determination the change in colour is very faint and difficult to distinguish. For better detection of end point, Yadav and Khera (1993) suggested not to add more than 25 mg of murexide indicator since, otherwise, the colour distinction will be very difficult. They also recommended to keep an untitrated sample, after addition of indicator, by the side of the sample being titrated to distinguish the colour change more precisely.

Calcium may also be determined with the help of a flame photometer by following the methods described for sodium and potassium and using calcium filter.

2.2–6 Hardness

Hardness of water is caused principally by the elements calcium (Ca) and Magnesium (Mg) and sometimes by iron and aluminium (Theroux et. al, 1943). Since iron and aluminium is seldom present in substantial amount in most of the pond waters, determination of Ca + Mg usually provides a fair idea of hardness of water and is expressed in terms of calcium carbonate ($CaCO_3$).

Principle

For the purpose of determining hardness of waters, concentration of Ca and Mg in water are first determined. The values are then converted to respective equivalents of $CaCO_3$ and added to get total hardness of water.

Requisites

Same as sec 2.2-5, in case of determination of Ca and Mg.

Procedure

Determine concentrations of Ca and Mg in water samples by following the procedure described in Sec. 2.2–5.

Calculation

(a) Observed ppm of Ca as $CaCO_3$ equivalent=ppm of Ca in water × Eq. weight of $CaCO_3$/Eq. weight of Ca = ppm of Ca in water × 50. 04 / 20.04.

(b) Observed ppm of Mg as $CaCO_3$ equivalent = ppm of Mg in water × Eq. weight of $CaCO_3$/Eq. weight of Mg = ppm of Mg in water × 50.04/12.16.

Total hardness of water = (a) + (b) ppm as $CaCO_3$.

Comments

Pond waters may be divided into various categories viz. soft, hard, very hard etc depending on their hardness (Sawyer and McCarthy, 1967). However, such classifications have little relevance to biological productivity of fish ponds. It is usually believed that the ideal range for hardness of pond water is 50 to 100 ppm.

2.2-7 Potassium

Fish ponds are usually considered to be fairly well supplied with potassium (Mandal and Chattopadhyay, 1992). Hence estimation of K has not so far received much attention in productivity assessment programmes of various aquaculture systems. However, with the intensification of fish culture practices and increasing uses of N and P fertilisers, K is gradually becoming the limiting nutrient element and understanding the K status of pond water is becoming an important proposition for judicious nutrient management in aquaculture.

Principle

K can be estimated gravimetrically, volumetrically, colorimetrically, potentiometrically or turbidimetrically (Yadav and Khera, 1993). However, in recent years, flame photometric analysis of K has become highly popular. Flame photometry involves emission of typically coloured flames from the basic ions, intensity of which are measured through a galvanometer. It is a rapid method and can be carried out with small amount of water sample.

Requisites

(a) Apparatus : Flame photometer with fuel gas supply.

(b) Glass wares : 50 ml beakers. 1ℓ and 100 ml volumetric flasks.

(c) Reagents : Standard K solution : Weigh 191 mg of AR grade potassium chloride (KCl) and dissolve in DW. Make the volume to 1 ℓ. This is 100 ppm stock solution of K. Take 10 ml of 100 ppm solution in a 100 ml volumetric flask and make up the volume to get 10 ppm K solution.

Procedure

Filter about 20 ml of water sample in a clean 50 ml beaker. Operate the flame photometer by igniting the flame, maintaining proper ratio of fuel gas and compressor air, as per the instructions of the instrument manual. Take some DW in a 50 ml beaker and atomise it through the atomiser. Adjust galvanometer reading to zero with DW. Then feed the 10 ppm K solution and adjust the galvanometer reading to 100. Thus, after adjustment,each unit (one) reading of galvanometer will be equivalent to 0.1 ppm of K. Now atomise the unknown water sample into the flame and record the reading.

Calculation

Concentration (ppm) of K in water=Reading $\times$ 0.1.

Comments

(a) The strength of standard solution and adjustment of galvanometer reading may be suitably modified as per the expected concentration of K in water.

(b) During estimations of large numbers of water samples, it is wise to check the galvanometer reading with the standard solution time to time. However, now a days many instruments are having in built automatic check knob.

Levels of K in Pond Water

Little information is so far available on occurrence of K in fish pond water and its relevance to productivity. However, Boyd (1978) observed that a concentration of 1.3 ppm may be considered to be the critical level of K in freshwater ponds.

2.2–8 Sodium

Generally, sodium (Na) is not included in regular analyses of pond water. However, in brackishwater ponds Na occurs in very high concentrations and determination of this element may be felt necessary. This element may also be present in higher amount in areas with alkaline soils.

Principle

As stated in case of K, flame photometric analysis is a very popular and easily operable method of determination of water-soluble Na. Principle of flame photometry has been discussed in sec. 2.2–7.

Requisites

(a) Apparatus : Flame photometer with fuel gas supply.

(b) Glass wares : 50 ml beakers, 1 ℓ and 100 ml vol. flasks.

(c) Reagents : Standard Na solution: Weight 0.254 g of AR grade dried sodium chloride (NaCl) and dissolve in DW. Make the volume of the solution to 1 ℓ. This is 100 ppm stock solution for Na. Take 10 ml of 100 ppm solution in a 100 ml vol. flask and make up the volume to 100 ml to get 10 ppm Na solution.

Procedure

Filter about 20 ml of water sample in a clean 50 ml beaker. Operate the flame photometer in the same manner as stated in sec 2.2-7 and adjust the reading of the flame photometer in such a manner so that each unit (reading) of the instrument becomes equivalent to 0.1 ppm of Na. Now use the water sample under study and record the reading.

Calculation

Concentration (ppm) of Na in water = Reading $\times$ 0.1

Comments

Effect of Na is usually better understood when studied with relation to other cations.

The proportion of Na ions in water with relation to total cation concentration is termed as soluble sodium percentage (SSP).

$$SSP = \frac{\text{Conc. of soluble Na (meq / } \ell)}{\text{Total cation conc. (meq / } \ell)} \times 100$$

For practical purposes, total cation concentration includes concentrations of Ca + Mg + Na + K, expressed in meq/1 (ppm/ equivalent wt for each cation).

Sodium absorption ratio (SAR) is a term which is used to express the relative concentration of Na ions over divalent cations viz. Ca and Mg.

$$SAR = \frac{\text{Conc. of } Na^{+} \text{ in meq / } \ell}{\sqrt{\text{Conc. (meq / } \ell) \text{ of } \frac{Ca^{++} + Mg^{++}}{2}}}$$

2.2-9 Alkalinity

The amount of acid required to titrate the bases in water is a measure of alkalinity (Boyd, 1978). Numbers of bases like hydroxides, carbonates, biocarbonates, ammonia, silicate, phosphate etc. may contribute to alkalinity of water. However, most of these bases usually occur in very small concentrations in general water bodies to exhibit considerable influence on water qualities. Hence, for practical purposes, presence of hydroxide (OH^-), carbonate ($CO_3^{=}$) and bicarbonate (HCO_3^-) are usually considered for determination of alkalinity.

Principle

Water samples containing measurable OH^- and /or $CO_3^{=}$ ions turn pink to phenolphthalein indicator. This water becomes colourless at pH below 8.4, when titrated with acid to convert these ions to HCO_3^- form. Again, water sample with only HCO_3^- can be titrated to the critical pH level of 5.3 with acid by using yellow to faint orange colour change of methyl orange indicator. Both of the indicators are used for determining total alkalinity, expressed as ppm of calcium carbonate ($CaCO_3$) and also the concentrations of HO^-, $CO_3^{=}$ and HCO_3^- ions in water samples.

Requisites

(a) Glass wares etc : 1 ℓ volumetric flasks, 250 ml conical flasks, burette, 100 ml cylinder etc.

(b) Reagents.

(i) Phenolpthalein indicator: Dissolve 0.5 g phenoloph-thalein in 50 ml of 95 percent ethyl alcohol and add to it 50 ml DW.

(ii) Methyl orange indicator (MO). Dissolve 0.5 g mehtyl orange in 100 ml DW.

(iii) Standard (0.02 N) Sulphuric acid (H_2SO_4). Prepare 0.02 NH_2SO_4 following the method described in appendix.

Procedure

Take 100 ml water sample in a 250ml conical flask and add 3 drops of phenolphathlein indicator. If the water turns pink, titrate with 0.02 N H_2SO_4 untill the pink colour just disappears. Note the volume (ml) of acid used. Then add 3 drops of MO indicator in the same (used) water sample. If the water turns yellow, titrate with same acid untill a faint orange end point is obtained. Note the volume (ml) of the acid used during this titration.

Calculation

If the volume (ml) of 0.02 N H_2SO_4 used for titration with phenolphthalein is P and the total volume of 0.02 N H_2SO_4 consumed during titration with both phenolphthalein and methyl orange is M, total alkalinity as ppm of $CaCO_3$ will be M × 10.

Theroux et al (1943) described following conditions for the purposes of estimation of concentrations (ppm) of OH_3^-, $CO_3^=$ and HCO_3^- ions in water in terms of $CaCO_3$ equivalent.

(i) if $P = M$, $OH^- = P \times 10$

(ii) if $P > \frac{1}{2}M$ $OH^- = (2P - M) \times 10$

$CO_3^= = 2(M - P) \times 10$

(iii) if $P = \frac{1}{2}M$, $CO_3^= = M \times 10$

(iv) if $P < \frac{1}{2}M$, $CO_3^= = 2P \times 10$

$HCO_3^- = (M - 2P) \times 10$

(v) if $P = O$, $HCO_3^- = M \times 10$

Comments

The uptake or release of carbon dioxide by organisms may change the proportions of $CO_3^=$ and HCO_3^- ions in water (Boyd, 1984). Hence water samples should be analysed as early as possible after collection, if forms of alkalinity are of major interest. Samples for total alkalinity may, however, be stored for a few days without appreciable changes.

2.2 -10 Chloride

Concentration of this ion varies widely depending on the environmental conditions. Most of the water soluble salts in a pond environment generally remains in Cl^- form and hence the amount of Cl^- in a pond water indicates very closely the total amount of soluble salts present in the ecosystem.

Principle

Cl^- ions can be conveniently estimated by titration with silver nitrate ($AgNO_3$) in presence of chromate ions. $AgNO_3$ forms silver chloride (AgCl) by reacting with the ($CrO_4^=$) ions present in water. When the Cl^- in water becomes exhausted, $AgNO_3$ then reacts with the $CrO_4^=$ ions to show a red colour of silver chromate ($AgCro_4$) indicating that the titration has been completed.

Requisites

(a) Glass wares, Burette, Pipette, Volumetric flasks, Porcelin basin etc.

(b) Reagents

(i) Potassium Chromate (K_2CrO_4) indicator : Dissolve 5 g of K_2CrO_4 in about 75 ml of water and add dropwise a saturated solution of $AgNO_3$ to form a red precipitate completely. Filter the suspension and dilute the filtrate to 100 ml.

(ii) Standard Silver Nitrate ($AgNO_3$) Solution: Weigh exactly 4.791 g $AgNO_3$ and dissolve in 1 ℓ of DW. This will produce a solution 1 ml of which is equivalent to 1 mg of Cl^-. Standardise the strength of the solution by titrating against a standard sodium chloride (NaCl) solution, by following the procedure part of this section.

(iii) Standard Sodium Chloride (NaCl) solution : Dissolve exactly weighed 1.648 g dry, analytical grade NaCl in 1 l of DW. 1 ml of this solution will contain 1 mg of Cl^-

Procedure

(a) Standardisation of $AgNO_3$ solution:

Take 10 ml of standard NaCl solution in a porcelin basin. Add DW to make the volume around 100 ml. Add 0.5 ml of K_2CrO_4 to the solution and titrate with $AgNO_3$ till appearence of a permanent red precipitation. Standardise the factor of the strength of $AgNO_3$ solution with the help of titre value.

If F be the factor of the $AgNO_3$ concentration and V the volume required for titration then $F = {}^{10}/_{V}$.

(b) Cl^- estimation :

Take 20 -100 ml of water sample in a porcelin basin, add DW, if necessary, to increase the volume and proceed as above to get the end point at formation of permanent red precipitation.

Calculation

If X be the titre value (ml) of $AgNO_3$, F the factor and V the volume of water sample (ml).

$$\text{Concentration of } Cl^-(\text{ppm}) = \frac{X \times F \times 1000}{V}$$

Comments

During titration of water sample with $AgNO_3$, the permanent red coloured precipitation can be visualised only after the reaction between Cl^- and $AgNO_3$ has been completed. Hence, for getting more accurate titre value, Theroux *et al.* (1943) suggested that a value of 0.2 may be deducted from the observed burette reading and then used for calculation.

Levels of Cl^- in pond water

Detailed studies on optimum concentration of Cl^- ions in

freshwater aquaculture are lacking. However, 1- 100 ppm concentrations are usually considered to be favourable.

2.2 -11 Sulphate

Sulphate ($SO_4^=$) is not commonly included in regular analysis of pond water. However, in some occasions this ion may occur in quite large concentrations and the necessity of estimating this ion may be felt.

Principle

$SO_4^=$ ions are widely estimated turbidimetrically after formation of barium sulphate ($BaSO_4$) through reaction of this anion with barium chloride ($BaCl_2$, $2H_20$) and making the colloid stable at pH 4.8 by adding sodium acetate - acetic acid buffer.

Requisites

(a) Apparatus — colorimeter.

(b) Glasswares — Volumetric flasks 1 *l*., 25 ml pipette etc.

(c) Reagents :

(i) Standard $SO_4^=$ solution : Weigh accurately 0.2719 g of analytical quality potassium sulphate (K_2SO_4). Dissolve in DW and make up the total volume to 1 *l*. This solution corresponds to 50 ppm of sulphur (S) in the form of $SO_4^=$.

(ii) Sodium Acetate - Acetic acid buffer (pH 4.8) : Dissolve 100 g sodium acetate in about 750 ml DW. To this add 31 ml of glacial acetic acid. Make up the volume to 1 *l*.

(iii) Gum acacia : Dissolve 2.5 g gum acacia in 1 *l* of DW keep for one night and filter.

(iv) Barium chloride ($BaCl_2$) powder : Available in the market. This should be less than 50 mesh size.

Procedure

(a) Preparation of standard curve : Take 0.5 ml, 1.0 ml, 2.5 ml, 5.0 ml and 10.0 ml of 50 ppm standard $SO_4^=$ solution in each of 25 ml volumetric flask respectively. Add a few ml of DW whenever required to make the total volume around 10 ml. Then add 1 ml of gum acacia and 1 g of $BaCl_2$ powder to each volumetric flask. Make the volume to 25 ml with DW and shake the contents thoroughly. Read the absorbance of the contents of each of the flask colorimetrically at 440 nm. Prepare standard curve for sulphur (S) as $SO_4^=$ using the concentrations of $SO_4^=$ in each of the volumetric flasks and their respective absorbance values.

Conc. of S in 25 ml volume

0.5 ml	—	1.0 ppm
1.0 ml	—	2.0 ppm
2.5 ml	—	5.0 ppm
5.0 ml	—	10.0 ppm
10.0 ml	—	20.0 ppm

(b) Estimation of S and $SO_4^=$

Take 10 ml of water sample in a 25 ml volumetric flask, add 10 ml sodium acetate-acetic acid buffer and mix thoroughly. Then add 1 ml gum acacia solution and 1 g $BaCl_2$ powder and shake again. Read the absorbance of this flocculated solution in a colorimeter at 440 nm as stated earlier. Compare the obtained absorbance with the standard curve to get the ppm concentration of S in the solution of volumetric flask.

Calculations

(a) Concentration of S in the water sample present in $SO_4^=$ (ppm) form = ppm reading in standard curve × 2.5.

(b) Concentration of S (ppm) × 3 = Concentration of SO_4 ions (ppm).

Comments

Turbidity of the samples can be measured more effectively with the help of a nephalometer or turbidity meter.

2.2-12 Dissolved inorganic nitrogen

Role of nitrogen (N) in fish nutrition is well documented

(Boyd 1978; Neess, 1949 and others). Inorganic forms of N dissolved in water are present predominantly in ammonium (NH_4^+) and nitrate (NO_3^-) forms excepting in highly polluted water where nitrite (NO_2^-) form may also occur in considerable concentration. Both of these NH_4^+ and NO_3^- forms are readily available to primary fish food organisms. Estimation of these two forms of N, therefore, constitutes a measure of availability of N in pond water - while determination of nitrite is often carried out in polluted waters.

A. Ammonium and nitrate forms

Principle

To determine the amount of NH_4^+ and NO_3^- forms of N in water, NH_4^+ - N is first distilled with alkali to release ammonia (NH_3) which is absorbed in an acid. Then NO_3^- form of N is reduced to NH_4^+ with hydrogen (H_2) in alkaline solution and the produced NH_4^+ is estimated as before. In the present method, the released NH_3 is absorbed in boric acid (H_3BO_3) to form NH_4 borate. This NH_4 borate is titrated back to original H_3BO_3 with a standard acid and the concentration of NH_4^+ is determined from the amount of standard acid required for this titration.

Requisites

(a) Apparatus : Kjeldahl distillation set with condenser.

(b) Glass wares : Kjeldahl flask, 250 ml conical flask, 1 *l* beaker, 1 *l* Vol. flask, pipette, burette etc.

(c) Reagents :

  (i) 40 per cent Sodium Hydroxide (NaOH) solution. Dissolve 400 g NaOH in a 1 *l*. beaker using about 750 ml DW, allow the solution to be cooled and make up the volume to 1 *l*.

  (ii) Davarda's alloy (Cu 50; Al 45 and Zn 5 per cent each): Available in market.

  (iii) Mixed indicator : Dissolve 0.5 g bromo cresol green and 0.1 g methyl red in 100 ml 95 per cent ethyl alcohol. Adjust the solution to bluish purple mid colour (pH 4.5) with diltute HCl or NaOH.

(iv) 4 per cent H_3BO_3 solution: Dissolve 40 g H_3BO_3 in about 750 ml DW, add 5 ml of mixed indicator and make up the volume to 1 ℓ. This stock solution should be adjusted by gradual addition of diltute HCl or H_2SO_4 untill the bluish colour changes to pink.

(v) Standard (0.02 N) H_2SO_4:

Prepare the reagent by following the method described in Appendix - 1.

Procedure

(a) $NH_4^+ - N$

Take 250 ml of filtered water sample in a kjeldahl flask and fit the flask with the distillation set alongwith condenser. Add 20 ml of 4 per cent H_3BO_3 in a 250 ml conical flask and put it below the outlet of the condenser so that the tip of the outlet dips into the H_3BO_3 solution. Add slowly 10 ml of 40 per cent NaOH in the Kjeldahl flask and fit the mouth of the condenser immediately. Carry out distillation till about 40 ml of distillate is accumulated in the receiving flask containing H_3BO_3. Stop heating and take out the conical flask. Titrate the distillate with 0.02 N H_2SO_4 till a pinkish colour appears.

(b) $NO_3^- - N$

Same (residual) water sample after NH_4^+ estimation can be used for determination of NO_3^--N. Take 20 ml 4 per cent H_3BO_3 in another conical flask and place beneath the condenser tip as before. Add about 0.5 g Davarda's alloy to the water sample, close the mouth of the Kjeldahl flask immediately and continue distillation to near dryness of water sample. Titrate the NH_4 borate in the conical flask as stated earlier.

(c) Total Inorganic ($NH_4^+ + NO_3^-$) N

In case separate estimations of NH_4^+ and NH_3^- forms are not required and it is required to determine the total dissolved inorganic N ($NH_4^+ + NO_3^-$) only, 250 ml of filtered water sample may be straightway distilled after addition of Davarda's alloy and 10 per cent NaOH to almost dryness. The liberated NH_4^+ is absorbed in 4 per cent H_3BO_3 and determined as stated above.

Calculations

N (ppm) in the form of NH_4^+ or NO_3^- or $NH_4^+ + NO_3^- = \frac{A}{V} \times 280$ where A = ml of 0.02 N H_2SO_4 required for titration in respective case. V = Volume (ml) of water sample used.

Alternative methods

NH_4^+ and NO_3^- forms of N can also be determined by using colorimetric methods using Nessler's reagent and phenol disulfonic acid respectively. However, in case of presence of considerable amount of salts in water, both of these methods may suffer from interference.

Comments

(a) Neither the volume nor the strength of H_3BO_3 need to be known exactly because it is the NH_4 borate formed that is titrated back to H_3BO_3 by H_2SO_4 (Jackson, 1971).

(b) To avoid bumping and frothing, add a few glass beads and a little amount of liquid paraffin to the water sample.

(c) Nitrite (NO_2^-) ions are usually present in very small amount in ordinary water bodies and they also become reduced and included alongwith NO_3^-. In case such inclusion is to be avoided, 1ml of aqueous sulfamic acid may be applied to the flask before addition of Davarda's alloy (Yadav and Khera, 1993).

Levels in pond water

Water soluble inorganic N ($NH_4^+ + NO_3^-$) below 0.1 ppm has been considered as indication of poor pródoctivity of pond and values above 0.2 ppm of good productivity by Banerjea (1967) under Indian condition.

B. Nitrite form

Principle

In a strongly acid medium, HNO_2^-, reacts with sulphanilamide to form a diazonium compound which reacts quantitatively with N - (1 - naphthyl) ethylenediamine dihydrochloride to form a strongly coloured azo compound (Golterman et. al, 1978). The intensity of this colour can be measured colorimetrically and the concentration of NO_2^- ion in the sample be determined.

Requisites

(a) Apparatus : Colorimeter or spectrophotometer

(b) Glass wares : 1*l*, 100 ml volumetric flasks, pipettes etc.

(c) Reagents :

(i) 6 N Hydrochloric acid (HCl) : Dilute AR quality HCl with equal amount of DW to get approximately 6 N HCl.

(ii) 0.2 per cent Sulphanilamide : Dissolve 0.2 g of the chemical in 100 ml DW.

(iii) 0.1 per cent N (1-naphthyl) ethylenediamine dihydrochloride: Dissolve 0.1 g of the reagent in 100 ml DW. Store in a brown bottle.

(iv) 5 per cent Ammonium sulphamate ($NH_2\ SO_3\ NH_4$) : Dissolve 5 g of the reagent in 100 ml DW.

(v) 5 N Sodium hydroxide (NaOH) : Dissolve 200 g NaOH in 1*l* DW.

(vi) Standard nitrate solution (1000 ppm) : Dissolve 6.08 g of AR quality potassium nitrite (KNO_2), dried at 105°C, in DW. Add 1 ml 5 N NaOH and dilute to exactly 1 *l*.

Dilute this solution 50 times to get 20 ppm standard solution of NO_2^- N.

Procedure

Take 90 ml of water sample in a 100 ml volumetric flask, add 5 ml sulphanilamide solution and 2 ml 6 N HCI and mix thoroughly. Allow the mixture to settle for 3 minutes, add 1 ml of 5 per cent ammonium sulphamate, mix, wait for another 3 minutes and add 1 ml napthylethylene diamine solution. Make up the total volume to 100 ml with DW. Wait for 15 minutes and read the absorbance of the solution at 530 nm against a blank solution.

Prepare a standard curve by using different concentrations of the standard solution and following the same procedure for estimation of NO_2^-. Use 2 ml, 5 ml, 10ml, 20 ml and 50 ml of the 20 ppm standard NO_2^- - N solution in 100 ml volumetric flasks and proceed as stated earlier. Under final volume of 100 ml, these solutions will have 0.4, 1.0, 2.0, 4.0 and 10.0 ppm of NO_2^-N respectively.

Calculation

If observed concentration of NO_2^- N of the water sample, as compared to the standard curve, is x ppm, actual concentration of NO_2^- N in the sample will be x X 1. 11 ppm.

Comments

NO_2^- is very unstable in nature and should be analysed immediately after collection. In case it is not possible to carry out the analysis immediately, the sample may be treated with 2 ml 6 N HCl and 5 ml sulphanilamide solution and preserved for about 24 hrs in refrigerator without much problem.

Levels in pond water

Higher concentration of NO_2^- is toxic. Although detailed study in this aspect is lacking, concentration below 0.2 ppm of water-soluble NO_2^- N is usually considered to be safe for aquaculture.

2 . 2 –13 Phosphorus

Phosphorus (P) is considered as the most critical nutrient element in the maintenance of pond productivity (Jhingran, 1975). Owing to its utmost importance to primary fish food organisms and extremely low availability in most of the aquatic ecosystems, this element has a special significance in fish culture.

Principle

In natural waters P occurs mostly as phosphate. This form of P is widely estimated colorimetrically after development of phosphomolybdic blue colour (Dickman and Bray, 1940). Stannous chloride ($SnCl_2$) has long been used as the reducing agent for developing the blue colour. But the colour developed by $SnCl_2$ doses not stay long and hence in recent years ascorbic acid is being used in stead of $SnCl_2$ (Murphy and Riley, 1962). The blue colour developed with ascorbic acid is stable for about 2 hrs.

Requisites

(a) Apparatus: Colorimeter or Spectrophotometer.

(b) Glass wares : Vol. flasks 100 ml, 25 ml, pipette etc.

(c) Reagents :

(i) Standard P solution : Dissolve 0.2195 g anhydrous potassium dihydrogen orthophosphate (KH_2PO_4) in about 500 ml of DW, add 5 ml conc. H_2SO_4, make up the volume to 1 *l*. This forms 50 ppm solution of P. 4 ml of this 50 ppm solution is to be diluted to 100 ml to get a 2 ppm solution of P which is used for making standard curve for P.

(ii) Ammonium molybdate - Potassium antimony tartarate solution :

Dissolve 1.0 g ammonium molybdate $(NH_4)_6 Mo_7 O_{24}$, 4 H_2O) and 0.02 g potassium antimony tartarate in about 500 ml DW. Add slowly 15 ml of conc. H_2SO_4 to this solution, make up the volume to 1 l. with DW.

(iii) Ascorbic acid - molybdate solution :

Weigh 88 mg ascorbic acid and add to 100 ml of the molybdate tartarate solution. This solution should be freshly prepared before P estimation.

Procedure

(a) Prepare a standard curve with different concentrations of P from the 2 ppm solution. Following concentration of P may be used for developing colour in 25 ml volumetric flasks.

Table 2 : Concentrations of P solution for making standard curve

ml of 2 ppm P soln.	*Conc. (ppm) of P in 25 ml vol. flask*
0.25	0.02
0.5	0.04
1.25	0.1
2.5	0.2
5.0	0.4

Take the above mentioned volumes of 2 ppm P solution in 5 nos of 25 ml volumetric flasks and also run a blank in another flask without any P solution. Add some amount of DW in each of the flasks so that the total volume comes up to 10-12 ml. Add 5 ml of ascorbic acid-molybdate mixture to each of the flasks. Make up the volume to 25 ml with DW and wait for 30 minutes for full development of blue colour. Read the optical density in a colorimeter or spectrophotometer at 660 nm wave length after setting O optical density (OD) with the blank solution. Prepare a standard curve by plotting concentrations (ppm) of P in X axis and respective optical densities in Y axis.

(b) Take 5-10 ml of water samples in 25 ml volumetric flasks depending on expected concentrations of P in waters. Also take a blank flask as before. Add adequate amount of DW to each flask and 5 ml of ascorbic acid-molybdate mixture. Make up the volume to 25 ml and determine the OD of the developed blue colour as before.

Put the OD values of unknown water samples on the standard curve to get the concentrations of P in the volumetric flasks as stated earlier.

Calculations

If the concentration of P in 25 ml volumetric flask is X ppm and the initial volume of water sample taken in the flask is V ml,

$$\text{concentration (ppm) of P in water sample} = \frac{X \times 25}{V}$$

Comments

(a) While estimating P, strict attention should be paid to keep the glass wares and reagents completely free from P because very minute concentration of P may interfere with P estimation.

Use of double DW gives better results.

(b) Sometimes P_2O_5 is used to denote the concentration of P in water and soil. Approximate relationship between P_2O_5 and P is $P \times 2.3 = P_2O_5$.

Levels of P in Pond Water

Benerjea (1967) studied concentrations of P in numbers of fish pond waters under Indian condition with relation to their productivity. He suggested that > 2.0 ppm of P_2O_5 in pond water may be considered to be optimum while 0.05 - 2.0 ppm may be considered to be moderate and < 0.05 ppm to be poor.

2.2-14 Dissolved Oxygen

Dissolved Oxygen (DO) is probably the most widely analysed chemical parameter in fish pond water quality assessment. Apart from its direct effect on respiration of different biotic organisms, it controls directly or indirectly many other limnological properties of pond and thus governs the productivity of the ecosystem to a great extent. Two most popular methods of DO estimation are - i) Winkler's method, and ii) Polarographic technique.

A. Winkler's method

This method (Winkler, 1888) is the most widely adopted technique for estimation of DO and has been in use throughout the world since long time.

Principle

Manganous (Mn^{2+}) ions form manganous hydroxide ($MnOH_2$) under highly alkaline condition. This $MnOH_2$ reacts with DO of water and an equivalent amount of Mn^{2+} is oxidised to higher valency states. On acidification, in presence of iodide, the oxidised manganous (Mn^{4+}) again reverts to divalent Mn^{2+} and iodine (I_2) is liberated in equivalent amount of DO. This I_2, estimated by titration with sodium thiosulphate ($Na_2S_2O_3$) indicates the amount of DO in water sample.

Requisites

(a) Glass wares : 1 ℓ volumetric flasks, pipette, burette, 100 ml conical flask, 125 ml bottle etc.

(b) Reagents :

(i) Manganous Sulphate ($MnSO_4$) : Dissolve in about 750 ml of distilled water (DW)

	480 g	$MnSO_4$, $4H_2O$
or	400 g	$MnSO_4$, $2H_2O$
or	365 g	$MnSO_4$, H_2O

boil, if necessary, cool, filter and make the volume to 1*l*.

(ii) Alkaline Iodide: Dissolve 500 g sodium hydroxide (NaOH) and 150 g potassium iodide (KI) in about 350 ml of DW separately. Allow the NaOH solution to cool down and mix the two solutions. Dissolve 10 g sodium azide (NaN_3) in 40 ml DW and add to the previous solution. Make up the final volume to 1*l*.

(iii) Standard Sodium Thiosulphate ($Na_2S_2O_3$) :

To prepare 0.1 N $Na_2S_2O_3$ stock solution, dissolve exactly 24.82 g of crystalline $Na_2S_2O_3$, $5H_2O$ in 700 ml of DW. Add a pellet of sodium hydroxide (NaOH) or 1 g sodium carbonate (Na_2CO_3) and dilute to exactly 1*l*. This stock solution should be stored in a brown bottle preferably below 25°C and standardised before preparing 0.025 N $Na_2S_2O_3$ solution by following the procedure mentioned below.

Dry crystalline potassium dichromate ($K_2Cr_2O_7$) in an oven at 125°C, cool and weigh accurately 4.904 g of the reagent. Dissolve it in DW and make the volume exactly 1 *l*. This gives 0.1 N solution of $K_2Cr_2O_7$. Take 25 ml of this solution in a 250 ml conical flask, add 1 ml of alkaline iodide and acidify the solution with 2 ml concentrated H_2SO_4. Titrate the liberated iodine with 0.1 N $Na_2S_2O_3$ solution using starch as indicator during later part of the titration. Application of starch turns the solution blue and it suddenly becomes colourless when the end point is achieved. Find out the strength of the $Na_2S_2O_3$ solution with the help of the titre value.

$$\text{Strength of } Na_2S_2O_3 = \frac{\text{Vol. of } K_2Cr_2O_7 \text{ used}}{\text{Vol. of } Na_2S_2O_3 \text{ required}}$$

Dilute required amount of stock solution to get 250 ml or any other volume of 0.025 N $Na_2S_2O_3$ by using the relationship.

$$V_1S_1 = V_2S_2$$

V_1 = Vol. of stock solution required to be diluted

S_1 = Strength of stock solution

V_2 = Desired vol. of solution after dilution (250 ml here)

S_2 = .025

(iv) Starch Indicator :

Add 1 g of starch powder to 10 ml DW, warm to about 80-90°C to dissolve in water, allow to cool down and add 0.1 g salicylic acid as preservative.

(v) Conc. Sulphuric Acid (H_2SO_4)

Procedure

Collect water sample in a 100 ml bottle. Add 1 ml each of $MnSO_4$ and KI solution through pipettes dipping slowly in the bottom of the bottle. Care should be taken to create minimum disturbances to the sample during these exercises. Close the bottle and invert it a few times to ensure thorough mixing of the water with the reagents. This will result in development of a flocculent precipitation which will gradually settle in the bottom. Whitish colour of the precipitation indicates poor status of DO while reddish to brown colour represents moderate to high values.

Add 1 ml of conc. H_2SO_4 and invert the stoppered bottle a few times to dissolve the precipitate. Take 50 ml of the solution to a 250 ml conical flask and titrate with freshly prepared 0.025 N $Na_2S_2O_3$ till the colour of the solution turns fade. Add 0.5 ml of starch indicator to form a blue colour and continue the titration carefully. At the end point, the blue colour will suddenly change to colourless.

Calculation

Concentration of DO in water sample (ppm) = ml of 0.025 N $Na_2S_2O_3$ used X4.

Comments

Presence of comparatively higher amount of oxidising or reducing substances interfere with this estimation procedure by reacting with iodide or iodine. Boyd (1978) described several methods for removing various interferences. Probably the most common interference is caused by nitrite (NO_2^-) ions which may be removed by application of sodium azide (NaN_3).

$$NaN_3 + H^+ = HN_3 + Na^+$$

$$HN_3 + NO_2^- + H^+ = N_2 + N_2O + H_2O$$

Considering the possibility of the problem with NO_2^- ions, NaN_3 application with alkaline iodide has been included in reagent preparation.

In some cases it may not be possible to complete DO estimation in situ. In such situations, the water sample should be treated with $MnSO_4$ and KI to form precipitation and then transferred to laboratory for subsequent analysis.

B. Polarographic method

Polarographic method of DO estimation ensures rapid and direct measurement of DO value in water. This method is rapidly becoming popular.

Principle

Polarographic oxygen analyser consists of a probe comprising of two electrodes both bathed in potassium chloride (KCl) and separated from the water sample by means of a membrane. The electrode produces a current proportional to the tension of oxygen in water and the instrument converts the current to a meter indicating DO concentration. Operation procedures of different DO analysers, although rather simple, vary in different instruments and depend on manufacturer's instructions. Hence they have not been dealt here.

Comments

It is desirable to check, time to time, the DO values obtained through DO analysers. This may be accomplished by comparing the DO values of same water sample by estimation through both Winkler's method and Polarography.

DO levels in water

Various observations have been made on critical as well as optimum levels of DO in pond water. Boyd (1978) stated that the minimum concentration of DO, tolerated by fish, is obviously a function of exposure time. A fish might survive 0.5 ppm DO for a few hours but not for several days. Further, the minimum tolerable concentration of DO will vary with species and size of fishes, physiological condition, concentrations of solutes and other factors. Ellis (1937) indicated that under average conditions, 3.0 ppm

DO or less should be regarded as hazardous for fish and for a good, varied fish fauna, the DO level should be 5.0 ppm or more. Mckee and Wolf (1963) stated that the DO of warm water fish habitat should not be less than 5 ppm during at least 16 h of any day. It may, however, be less than 5 ppm for a period less than 8 h but at no time DO should be less than 3.0 ppm. Swingle (1947) made a similar assessment (Table- 3) which has been widely accepted by different workers.

Table 3 : Effects of DO levels on fish population.

DO (ppm)	*Effects*
< 1	Lethal on exposure for more than a few hours.
> 1 — 5	Fish will survive but reproduction and growth will be poor.
> 5	Normal survival and growth.

2. 2–15. Free Carbon Di-oxide

Apart from its important role in photosynthesis of primary fish food organisms, free carbon di-oxide (CO_2) has interdependence with pH and bicarbonate - cabonate equilibrium. It also renders some essential nutrient elements in soluble form through formation of cabonic acid. However, if present in higher concentration, it may exert adverse effects on respiration and other physiological functions of biotic lives present in the pond ecosystem.

Principle

Waters having more than 8.3 pH does not contain appreciable amount of free CO_2. Therefore, phenolphthalein indicator, which has end point at this pH value is used for estimating the amount of CO_2 in a water sample for titration with a standard alkali.

$$2NaOH + CO_2 = Na_2CO_3 + H_2O$$

Requisites

(a) Glass wares : Burette, pipette, volumetric flasks, porcelin basin etc.

(b) Reagents.

 (i) Phenolphthalein indicotor

Dissovle 0.5 g of phenolphthalein in 50 ml of 95 per cent ethyl alcohal. Make up the volume to 100 ml with recently boiled DW.

(ii) N/44 Sodium Hydroxide (NaOH)

Prepare a N/10 stock solution of NaOH by dissolving exactly 4.0 g A.R. quality NaOH pellets in 1 l DW. Standardise the solution by titrating it against N/10 H_2SO_4 (see appendix-1) using phenolphthalein indicator untill a faint pink colour appears. Dilute 100 ml of this N/10 NaOH solution to 440 ml with DW to get N/44 NaOH solution. The prepared standard solution should not be used for a long period and while preparing fresh standard solutions, the stock solution should be standardised every time.

Procedure

Take 100 ml of the freshly collected water sample in a white porcelin basin. Add 3-4 drops of phenolphthalein indicator into the water. If the sample turns pink, the pH of water is above 8.3 and free CO_2 is not present. If the solution remains colourless after addition of indicator, titrate it with N/44 NaOH with gentle stirring with a glass rod till the colour turns pink.

Calculation

Concentration of free CO_2 (ppm) = ml of N/44 NaOH required for titration X 10.

Comments

Free CO_2 is very labile and hence water samples for free CO_2 estimation should be collected with caution so that it does not come in undue contact with atmospheric water. The same care should be exercised at the time of analysis also so that minimum changes occur in the water sample due to exposure to atmosphere. Boyd (1978) suggested that this analysis should be completed within 2–3 h of collection.

Since there is possibility of escape of free CO_2 from the porcelin basin due to stirring operations during titration, some workers suggest that the titraion should preferably be carried out in a Nessler's tube or glass cylinder using a white background.

Levels of CO_2 in water

Various critical levels of free CO_2 in aquatic habitat have been suggested by many workers. Ellis (1937) stated that the water supporting good fish population usually contains less than 5 ppm of free CO_2. Under intensive aquaculture, free CO_2 levels may fluctuate between O ppm during afternoon and around 5-10 ppm during dawn without any apparent ill effect on fishes (Parks et. al, 1975). However, the most detrimental influence of free CO_2 in fish culture may result when CO_2 concentration increases during critically low DO levels (Boyd, 1978). Unfortunately, this situation is observed in most of the cases of CO_2 toxicity.

2.2-16 Dissolved organic matter

Oxygen consumed during oxidation of organic matter may be considered as an index of dissolved organic matter in water. Usually potassium permanganate ($KMnO_4$) is used for this purpose and the value is some times referred as potassium permanganate demand of water also.

Principle

When $KMnO_4$ is used for oxidising soluble organic matter, the MnO_4 ion is reduced to MnO_2^- form and the pink colour becomes faint. This change in colour is used to determine the amount of oxygen required to oxidise the organic matter.

Requisites

(a) Glasswares: Volumetric flasks (1 ℓ), conical flasks (250 mi), pipette, burette etc.

(b) Reagents

  (i) Potassium permanganate ($KMnO_4$) solution: Dissolve exact 0.4 g $KMnO_4$ in 250 ml DW. Each ml of this solution will be equivalent to 0.1 mg O_2.

  (ii) Ammonium oxalate ($C_2H_8O_4N_2$). Dissolve 0.888 g $C_2H_8O_4N_2$ in 1ℓ DW.

  (iii) Dilute H_2SO_4: Add slowly and cautiously 100 ml conc. H_2SO_4 to 300 ml DW.

Procedure

Take 50 ml filtered water sample in a 250 ml conical flask, add 5 ml 1 : 3 H_2SO_4 and shake the flask gently. Add 10 ml of $KMnO_4$ to this acidified water and heat in a waterbath. After half an hour, remove the flask from the waterbath and add 10 ml NH_4 oxalate to it. This will result in disappearance of the existing pink colour of the solution. Now add to this colourless solution, $KMnO_4$ very slowly till the pink colour just reappears. Record the burette reading.

Calculations

Amount of O_2 (ppm) required to oxidise the dissolved
OM = ml of $KMnO_4$ required X 2.

Comments

In case of occurrence of high amount of dissolved organic matter e.g. in waste water ponds, the pink colour may disappear during heating itself. In that case the procedure should be repeated using higher amount of $KMnO_4$ solution followed by equal amount of NH_4 oxalate.

2.2-17 Biochemical Oxygen Demand

Demand for oxygen in a waterbody may be exerted by (a) carbonaceous organic materials and (b) oxidisable nitrogenous compounds which may react with dissolved molecular oxygen and thus create oxygen tension in the water (Banerjee and Chattopadhyay, 1980). The biochemical oxygen demand (BOD) is a measure of the amount of oxygen required by micro organisms to decompose the organic matter (OM) in a water sample under a specific set of conditions (Boyd, 1978). The property has a direct bearing on the oxygen balance of water, particularly when the environment is rich in organic materials. Complete decomposition of OM may require a long time and hence a 5 day incubation period has been suggested for most practical purposes (APHA, 1971). The BOD value thus obtained is usually termed as BOD_5.

Principle

For determination of BOD_5 value, an aliquot of the water

sample is incubated in dark at 20°C for 5 days and the DO level achieved in this sample is deducted from the initial DO value of water. This gives an estimate of the amount of oxygen required to decompose the OM in the water body during 5 day period under the specified condition. In case of water samples rich in OM, the consumption of oxygen may be so high that the DO level of water may be completely exhausted. In such cases, suitable dilutions of the water samples are done in order to obtain the BOD value.

Requisites

(a) Apparatus : BOD incubator

(b) Glass ware : 1l conical flask, BOD bottle, all glass wares for DO estimation

(c) Reagents : Use the chemicals for DO estimation

Procedure

Take about 1 l of water sample in a suitable container and shake it thoroughly to increase its DO level. Fill two BOD bottles (narrow neck glass stoppered bottle of usually 250 ml volume) with the water sample. One of these bottles should be black coloured or wrapped with black paper/polythene sheet to avoid entry of light. Incubate this opaque bottle in a BOD incubator at 20°C. Determine the DO of the other bottle (DO) by following the method stated in sec. 2.2-14. Bring out the incubated bottle from the incubator after 5 days and determine the DO of this water sample (DI).

If OM content of the water sample is so high that there is possibility of absolute depletion of DO value on incubation, the water sample should be diluted with DW at suitable ratios depending on approximate organic load of the water sample. Use this diluted water as the sample and proceed as stated above.

Calculations

BOD_5 (ppm) = DO (ppm) - DI (ppm)

In case of diluted sample - if the water has been diluted x times,

BOD_5 (ppm) = [DO (ppm) - DI (ppm)] X x

Comments

Boyd (1978) stated that standard BOD_5 value of water may be of little use in fish culture where momentary rate of consumption of DO by the micro organisms in the water is of more interest. He has, therefore, presented a short term BOD estimation procedure for unaltered water samples. Under this method, a water sample is incubated in pond water at desired depth for a few hours and the BOD is estimated from the differences in initial and final DO values.

Levels of BOD in pond water

Fish ponds generally exhibit BOD_5 values between 5–20 ppm. Higher BOD_5 value indicate occurrence of higher amount of organic matter in pond water and, thereby, possibility of oxygen depletion. However, pond waters having high BOD load usually contain more nutrient elements. Chattopadhyay et. al. (1988), while studying all these factors, suggested 10–20 ppm BOD_5 load of pond water to be optimum for aquaculture.

2.2-18 Chemical Oxygen Demand

The chemical oxygen demand (COD) is a measure of the total amount of oxygen which is required to oxidise all the organic matter in a sample to CO_2 and H_2O (Boyd, 1978). This property, therefore, indicates the organic matter concentrations of pond water and also exhibits close correlation with BOD values as well (Sawyer and McCarty, 1967).

Principle

In the measurement of COD, all the organic matter present in a water body are oxidised with the help of a standard oxidising agent. The amount of oxidising agent consumed during this process indicates the quantity of oxygen required for this purpose and from this COD is determined. In the method presented in this section, potassium dichromate ($K_2Cr_2O_7$) is used for carrying out the oxidation. The excess $K_2Cr_2O_7$ is back titrated by standard ferrous ammonium sulphate ($Fe(NH_4)_2(SO_4)_2$) solution to know the actual amount of standard $K_2Cr_2O_7$ required for oxidation of OM in the water sample. The equivalent requirement of oxygen for oxidising the same amount of OM as has been oxidised by $K_2Cr_2O_7$ is then worked out.

Requisites

(a) Apparatus : Refluxing system

(b) Glasswares : 250 ml conical flasks, burette, pipette, 1 *l* volumetric flasks etc.

(c) Reagents :

(i) 1 N Potassium dichromate ($K_2Cr_2O_7$) : Dissolve 49.036 g dry AR quality $K_2Cr_2O_7$ in about 750 ml of DW. Dilute it to exactly 1l. This serves as stock solution.

(ii) 0.025 N $K_2Cr_2O_7$: Dilute 25 ml of the stock (1N) solution to exactly 1 *l*.

(iii) 0.025 N Ferrous ammonium sulphate ($Fe(NH_4)_2(SO_4)_2$) : Dissolve 9.8 g of $Fe(NH_4)_2(SO_4)_2, 6H_2O$ in about 500 ml of DW, add about 20 ml of conc. H_2SO_4, cool the solution, make up the volume to 1 *l*. This reagent is easily decomposable and hence should be standardised with $K_2Cr_2O_7$ before use..

(iv) Diphenyl amine indicator : Dissolve 1 g of the indicator in 100 ml of conc. H_2SO_4.

(v) Orthophosphoric acid : Available in market.

(vi) Silver sulphate crystals : Available in market.

Procedure

Take 20 ml of water sample in a conical flask. Add a few glass beads. Mix exactly 10 ml of 0.025 N $K_2Cr_2O_7$ to the water sample and then add slowly 30 ml of conc. H_2SO_4. If the water is likely to contain large amount of chloride, e.g., in case of estuarine waters, add a few crystals of silver sulphate to the flask to overcome the interference of these ions. Attach the flask to a refluxing condenser and reflux for 2 hrs, over a heater. After refluxing wash the inside residues of the condenser into the flask, dilute with about 75-80 ml DW. Add 10 ml of orthophosphoric acid to prevent interference of ferrous ions during titration. Add about 1 ml of diphenyl amine indicator to develop a blue colour in the solution and titrate with standard Fe $(NH_4)_2(SO_4)_2$ untill the blue colour flushes to green. Run a blank in exactly the same procedure with all the reagents excepting the sample.

Calculation

Boyd (1978) suggested a simple calculation for determination of COD.

$$COD\ (ppm) = \frac{x(N) \times 8000}{V}$$

Where x =ml of Fe $(NH_4)_2$ $(SO_4)_2$ required in blank - ml of Fe $(NH_4)_2$ $(SO_4)_2$ required in sample titration.

N =Normality of Fe $(NH_4)_2$ $(SO_4)_2$ solution.

V =Sample volume.

Comments

In some cases the COD of water may be so high that 10 ml 0.025 N $K_2Cr_2O_7$ may not be sufficient to oxidise all of the OM present in it. In such cases higher amount or conc. of $K_2Cr_2O_7$ may be used.

Levels of COD in pond water

Like BOD, COD is also an index of the organic enrichment of pond water. However, since COD is measured by oxidising all the organic matter in the water sample, this value should be greater than BOD level of the same water body. Fish ponds generally have COD values of 10-200 ppm. However, the optimum range may be considered to be between 80 and 100 ppm.

2.2-19 Phenol

Phenols are aromatic compounds with hydroxyl groups within. Although phenols are known to offer resistance to disease and pest attack yet occurrence of phenol in higher concentration in water is an index of water pollution and estimation of phenols is frequently included in assessment of environmental condition of various water bodies.

Principle

Phenols produce characteristic molybdenum blue colour by reacting with phosho-molybdic acid in folin-ciocalteau reagent under alkaline condition. The intensity of this colour indicates the concentration of phenolic compounds in the medium.

Requisites

(a) Apparatus : Spectrophotometer/Colorimeter

(b) Glasswares : 100 ml conical flasks, porcelin basin, burette, pipette, graduated tu es etc.

(c) Reagents :

  (i) Ethyl alcohol

  (ii) Folin cio-calteu reagent- Available in market.

  (iii) 20 per cent Sodium Carbonate (Na_2CO_3) : Dissolve 20 g Na_2CO_3 in 100 ml DW.

  (iv) Standard catechol solution: Dissolve 0.1 g catechol in 100 ml DW. This will give 1000 ppm solution of catechol equivalent of phenol. Dilute 25 ppm of this solution to 250 ml for getting a working standard of 100 ppm.

Procedure

Take 0.5, 1.0, 2.5, 5.0 and 10.0 ml of 100 ppm catechol solution in 100 ml conical flasks and also run a blank without any phenol solution. Add 20 ml of ethyl alcohol to each. Shake for half an hour and keep overnight. On the next day, transfer the materials to porcelin basins and evaporate them to dryness through gentle heating. Cool the residue and dissolve it in exactly 5 ml DW and transfer to graduated tubes. Add 0.5 ml Folin cio-calteu reagent to each, shake well for a few minutes and add 3 ml 20 per cent Na_2CO_3. Make up the volume to 10 ml with DW, shake well and place the tubes in boiling water for exactly 1 min. Cool and read the optical density of the developed blue colour at 650 nm in a spectrophotometer or colorimeter against the blank.

Prepare standard curve using the concentrations of the standard solution under different aliquots.

Take 100 ml of water sample and proceed in the same manner, excepting using 100 ml ethyl alcohol instead of 20 ml. Make up the final volume to 10 ml, as stated earlier, and read the optical density (OD) of the developed colour. Match the observed OD reading with standard curve and find out the concentration of phenol in the final solution.

Table 4 : Concentrations of catechol equivalent of phenol under different dilutions

ml of standard 100 ppm soln. used	*Conc. (ppm) of phenol after final dilution*
0.5	5.0
1.0	10.0
2.5	25.0
5.0	50.0
10.0	100.0

Calculation

If observed concentration of phenol in final solution be x ppm, concentration in the original solution will be $\frac{x}{10}$ ppm.

Comments

Sometimes a whitish colour may be developed during boiling of the samples. In such cases the colour may be developed at 40 - 50°C, preferably using an incubator, giving more time for development of colour.

Since catechol is used for preparing the standard solution in this analysis, results should be more accurately expressed as catechol equivalent of phenols.

Levels of phenols in water

Very little work has been done on occurrence of phenolic compounds in fish pond water. However, under Indian Standard Units, occurrence of 0.5 ppm of phenol in water is considered to be safe limit for aquatic lives.

2.2-20 Hydrogen Sulphide

Hydrogen sulphide (H_2S) may be an important constituent of water quality under polluted condition which may affect the productivity of ponds considerably and be required to be determined under specific situation.

Principle

Hydrogen sulphide (H_2S) reacts with Iodine (I_2) in the following way

$H_2S + I_2 = 2HI + S$

This reaction is utilised for determination of H_2S by allowing sufficient amount of standard I_2 solution to react with H_2S in water and then titrating the excess I_2 with standard sodium thiosulphate ($Na_2S_2O_3$). The amount of I_2 reacted with H_2S is determined by subtraction from a blank value and the concentration of H_2S is then calculated.

Requisites

(a) Glasswares : Conical flasks (500 ml), 250 ml cylinder, burette, pipette etc.

(b) Reagents :

(i) 0.025 N Sodium thiosulphate ($Na_2S_2O_3$) : Prepare the reagent by following the method described in sec. 2.2-14).

(ii) Starch indicator : Preparation of this solution has also been stated in sec. 2.2-14.

(iii) 0.025 N Iodine (I_2) solution : Dissolve 2 g potassium iodide (KI) in small amount of boiled DW under hot condition. Cool the solution and add 3.173 g Iodine (I_2). Make up the volume to 1 l with DW. Titrate 25 ml of this solution with 0.025 N $Na_2S_2O_3$ using starch indicator following the method used for DO estimation (sec. 2.2-14). Adjust the strength of the I_2 solution so that 1 ml of this solution becomes equivalent to 1 ml of 0.025 N $Na_2S_2O_3$ solution.

Procedure

Take 10 ml of standard (0.025 N) iodine (I_2) solution in each of two conical flasks. Add a pinch of potassium iodide crystal to each. To one of the flasks, add very slowly 200 ml of the water sample and to the other 200 ml of DW. Titrate both the samples with 0.025 N $Na_2S_2O_3$ using starch as indicator till the blue colour of the solution becomes colourless, as stated in sec. 2.2-14.

Calculation

1 ml of 0.025 N I_2 solution is equivalent to 0.426 mg H_2S. Therfore, if the volume of I_2 reacted with H_2S in water sample be x ml (blank reading - sample titre value), concentration of H_2S in water sample will be $\frac{x \times 0.426 \times 1000}{200} = x \times 2.13$ ppm.

Comments

It is the un-ionised H_2S which is toxic for aquatic lives. Percentage of un-ionised H_2S usually tends to decline with increase in pH as shown in table 5.

Table 5 : Occurrence of un-ionised H_2S (%) in total sulphide content

pH	*Un-ionised H_2S (%)*
5.0	99.0
5.5	97.0
6.0	91.1
6.5	76.4
7.0	50.6
8.0	9.3
8.5	3.1
9.0	1.0

Levels of H_2S in water

Presence of even very small concentration of H_2S in pond water is undesirable. However, an amount of 0.05 ppm is generally considered to be critical.

2. 2-21 Chlorine

Free chlorine remains in very small concentrations in general pond waters. However, the amount may be higher under contaminated or polluted conditions and need to be estimated.

Principle

Free chlorine reacts with N - N diethyl- P-phenylene diamine (DPD) in presence of excess iodide to produce a red colour.

This colour disappears when the chlorine is reduced to chloride by ferrous ammonium sulphate (Santhanam et al, 1989). This reaction is utilised to determine the amount of free chlorine in pond water.

$$\underset{\text{red}}{\text{Chlorine — DPD}} + Fe^{2+} \rightarrow Cl^{-} + Fe^{3+} + \underset{\text{colourless}}{\text{DPD}}.$$

Requisites

(a) Glasswares : 250 ml conical flasks, pipettes, 250 ml, 50 ml, 1 l volumetric flasks.

(b) Reagents :

(i) DPD indicator: Dissolve 250 mg N-N-diethyl-P-Phenylene diamine (DPD) - oxalate in about 150 ml DW. Add 2 ml of 25 percent H_2SO_4 (1:3; conc. H_2SO_4 : DW). Make the volume to 250 ml with DW. This indicator should not be used for prolonged period.

(ii) Buffer solution : Dissolve 12 g disodium hydrogen phosphate (KH_2PO_4) in about 250 ml DW and add to the solution 10 mg mercuric chloride ($HgCl_2$). Dissolve 0.4 g ethylene diamine tetra acetic acid (EDTA) salt in 50 ml DW in a separate flask and add to the former mixture. Mix the contents and make up the volume to 1 l with DW.

(iii) Potassium iodide crystals : Available in market.

(iv) 1 N ferrous ammonium sulphate ($Fe(NH_4)_2(SO_4)_2$) : Dissolve 392.2 g of $Fe(NH_4)_2(SO_4)_2$ $6H_2O$ in 800 ml of DW, add 20 ml conc. H_2SO_4 and dilute to 1 l with DW.

Procedure

Mix 5 ml each of the buffer solution and the DPD indicator into a 250 ml conical flask. Add to this flask 100 ml of the water sample and 1 g of KI crystals. Mix thoroughly. Titrate the solution with 1 N $Fe(NH_4)_2(SO_4)_2$ untill the red colour just disappears.

Calculations

$$\text{Conc. of } Cl_2 \text{ (ppm)} = \text{Titre value (ml)} \times 354.5$$

Levels of Cl_2 in pond water

Free Cl_2 is extremely toxic to fish population. White (1955) observed only 0.028 to 0.08 ppm of Cl_2 to kill 100 per cent of fish population with varied species. Considering these facts Boyd (1978) suggested that occurrence of even measurable concentrations of Cl_2 in pond water should not be considered to be safe for holding fishes.

2.2–22 Temperature

Water temperature plays an important role in controlling most of the chemical as well as biological phenomena in a fish pond soil and, thereby, the productivity of the ecosystem.

Principle

For determining water temperature in fish pond water, a centigrade thermometer graduated at 0.1°C interval is usually used. If more accuracy is sought for, a thermometer with 0.01°C divisions may be selected. For subsurface water sampling, a specially designed reversing thermometer is used where the temperature reading can be made fixed after the thermometer comes in equili-brium with water temperature.

Requisites

Thermometer : Centigrade thermometer with divisions as per requirement

Procedure

For determining temperature of surface water, dip the thermometer directly into the water, keep it there for about one minute and note the temperature reading in the thermometer immediately.

In case of subsurface water samples, use the reversing thermometer. This thermometer is placed at the desired depth under water and, by the help of a trigger arrangement, the thermometer is reversed and the mercury reading is made fixed. The thermometer is taken out and the temperature reading is noted.

Comments

Very low or too high temperatures are usually considered to be noncongenial for fish culture. Generally 20-35°C range is considered to be optimum. However, this will depend on location of the pond, species of fishes etc. e.g. cold water fishes are adapted to grow well under very low temperatures.

2.2-23 Turbidity

Turbidity is a measure of the interference caused due to presence of suspended matters to the passage of light. Generally the water becomes turbid due to presence of organic matter as well as silt, clay, and other inorganic materials. Turbidity may also be caused by planktonic population in the pond waters.

Principle

Among different methods available for determination of turbidity, the method using Secchi disc is more commonly applied in various field studies and has been reported to produce good results. Now a days, turbidity meters are also being used in many laboratories for determination of turbidity, but in view of simplicity and less expense, the Secchi disc method has been discussed here.

Requisite

(a) Apparatus : Secchi disc - Available in market. May also be built by following the specification stated below.

Procedure

A standard Secchi disc is a circular metalic plate having 10 cm radius. The upper surface of the disc is divided into four quardrates, painted alternately in black and white colours. The disc is gradually lowered into the water with the help of a rope fixed to the centre of it and the depth at which the upper surface of the disc just disappears, is noted. Now the disc is lifted upward very slowly and the depth at which it reappears is noted.

Calculation

If the depth of water at which the disc disappears be d_1, and the depth at which it reappears be d_2, the value $\frac{d_1 + d_2}{2}$ in cm gives a measure of light penetration in the water body and is known as "Secchi disc transparency". An approximate idea about concentrations of suspended matter in water can be obtained from the transparency values by using table 6.

Table 6 : Relationship between transparency and suspended materials

1 *Transparency (cm)*	2 *Suspended materials (ppm SiO_2)*	1	2
2.0	1000	11.0	90
2.2	900	12.0	80
2.4	800	13.0	70
2.6	700	15.0	65
2.8	600	18.0	50
3.0	550	20.0	45
4.0	360	22.0	40
5.0	240	25.0	35
6.0	190		
7.0	145		
8.0	125		
9.0	110		
10.0	100		

Adopted from CICFRI Bull. no. 12

Comments

Wallen (1951) determined the direct effects of clay turbidity and observed behavioural reactions of fish to take place when turbidities exceeded 20,000 ppm. Most of the fishes survived exposure to 1,00,000 ppm of clay turbidity for a week or longer but appreciable mortalities occurred at turbidities above 1,75,000 ppm. However, Irwin (1945) reported that turbidity in natural waters seldom exceeded 20,000 ppm.

Abundance of fish food organisms may also create turbidity of pond water and may be considered desirable upto certain limit. It is generally believed that plankton turbidity of 20-40 cm is beneficial for aquaculture.

2.2-24 Alum requirement

Alum floculates and thus precipitates suspended clay particles from a waterbody and is widely recommended for controlling turbidity in fish ponds. However, alum has considerable effect on pH and alkalinity of water and hence the dose of application should be determined after giving due consideration to these effects.

Principle

Alum requirement (AR) for a waterbody can be determined by applying different doses of alum to the water sample and finding out the optimum dose with reference to changes in turbidity, pH and alkalinity. Efficiency of this method, however, will depend on the experience of the analyst than the method itself. Boyd (1979) described a simple method for determination of AR taking into consideration the changes in both pH and alkalinity of pond waters. The author has used this method in principle after making little modifications.

Requisites

(i) Apparatus : pH meter

(ii) Glasswares : Six nos. of 1 ℓ glass cylinders, 50 ml beakers, 1 ℓ volumetric flask etc.

(iii) Reagents

(a) Alum solution (5000 ppm) : Dissolve 5.0 g of alum (aluminium sulphate, $Al_2(SO_4)_3$) and dilute to 1 ℓ. distilled water. This solution should not be kept for long time.

(b) Buffer solutions for pH determination - as described in sec. 2.2-1.

Procedure

Take 1 ℓ water sample in each of six nos. of 1 ℓ glass cylinders. Use the first cylinder as control and add 1,2,3,4 and 5 ml of the 5000 ppm alum solution to the other cylinders respectively- so that they result in 5,10,15,20 and 25 ppm of alum concentration in the water samples. Keep the glass cylinders for 2 hr. After this period, find out the cylinder where maximum flocculation has occurred. Visual assessment is generally sufficient for this purpose.

In case there is any problem for visual observation, use a thin wooden rod with a bright pin fixed perpendicularly at one end of it. Dip this end of the stick slowly in the water of each of the cylinders and observe the visibility of the pin as the rod gradually goes down. Observe the depth of water at which the pin just disappears. Compare these values and find out the cylinder which exhibits maximum visibility.

Take out about 50 ml water in a beaker from the cylinder offering maximum transparency and determine the pH value of the water with the help of a pH meter as stated in sec. 2.2-1. If the change in pH is not very prominent, use the dose (ppm) of alum used in this cylinder as the AR of the water sample. On the other hand, if the pH of the water is decreased considerably say, below neutral level, use the water of the cylinder with next highest transparency and determine its pH. AR of the water sample will denote that dose of alum in water which will result in considerable flocculation of suspended materials and at the same time will not affect the pH of the water adversely.

Calculation

AR = Alum concentration in water (ppm) in the glass cylinder which exhibits good flocculation without adverse effect on pH.

Comments

If the stated doses of alum are not found to be suitable, lower or higher doses may be used depending on the situation.

Application of alum

It will be evident from the above discussion that application of alum reduces the alkalinity status of water. Hence this material should be used only in case of extreme necessity and should not be applied frequently.

2.2-25 Primary Productivity

Primary productivity (PP) of an ecological system, community or any part thereof, is defined as the rate at which radiant energy is stored by photosynthetic or chemosynthetic activity of producer organisms (chiefly green plants) in the form of organic substances which can be used as food materials (Odum, 1971). Of this, gross primary productivity (GPP) is the total rate of photosynthesis including the organic matter used up in respiration during the measurement period while net primary productivity (NPP) indicates the amount of organic matter that is stored in the plant tissues after meeting the demand of respiration. Close correlation between primary productivity and fish yield has been emphasized by many workers (Wolny and Grygierek, 1972; Sreenivasan, 1968 and others).

Principle

Primary productivity of a water body is usually determined from the differences in DO values of water samples incubated in bottles under light and dark conditions thus allowing photosynthetic activity to take place in one bottle and the same to be restricted under the other. The decrease in DO content in the dark bottle, as compared to initial value represents the amount of DO consumed by respiration by all the biomass in the bottle. The increase in DO in the light bottle indicates that amount of DO in water which exceeded oxygen consumption by respiration. Both GPP and NPP can be calculated from the differences in these DO values.

Requisites

(a) Glasswares.

(i) All the glasswares required for DO estimation (sec. 2.2-14).

(ii) Light and dark bottles : Use two 125 ml stoppered transparent glass bottles as initial and light bottles. For using as dark bottle, paint a similar bottle with black colour and then wrap with black cloth or polythene sheet in such a way that light penetration in the bottle can be prevented.

(b) Reagents : All the chemicals for DO estimation (Sec. 2.2-14).

Procedure

Take in water sample cautiously as well as uniformly in three bottles (initial, light and dark) from the desired depth of water. Add 1 ml of manganous sulfate and alkaline iodide each in the initial bottle. Invert the stoppered bottle a few times. Take the other two bottles and dip them in the water under the depth from which water samples were collected. Incubate the water samples under water for some period which should not normally be less than 3 hrs. After expiry of the time, take up both of the bottles and fix oxygen as before. Determine DO values of water in all the bottles as stated in sec. 2.2-14.

Calculation

$$\text{GPP} = \frac{\text{LB} - \text{DB}}{\text{T}} \times \frac{0.375}{1.2} \times 1000 \text{ mg cm}^{-3}\text{h}^{-1}$$

$$\text{NPP} = \frac{\text{LB} - \text{IB}}{\text{T}} \times \frac{0.375}{1.2} \times 1000 \text{ mg cm}^{-3}\text{h}^{-1}$$

Where—

LB	=	DO (ppm) in light bottle
DB	=	DO (ppm) in dark bottle
IB	=	DO (ppm) in initial bottle
T	=	Time (h) of incubation
0.375	=	ratio of the weight of C and O
1.2	=	photosynthetic co-efficient.

Comments

In some situations it may not be feasible to carry out incubation in the site of study itself. In such cases, the bottles, filled with water samples, may be brought to a convenient place and incubated under water at desired depth in a plastic pool or cistern etc. The transparency of this water should be as close as possible to that of pond water under study so as to enable the sunlight to penetrate the water body in a similar manner as would have been in case of pond water.

Levels of primary productivity in pond water

Considerable studies have been done on importance of primary productivity on production of fishes and it has been widely observed that fish production increases with increment in primary productivity. Good productive ponds usually exhibit GPP value of 2000 - 5000 mg C m^{-3} day^{-1}.

2.2-26 Chlorophyll-a

Several types of chlorophyll occur in plants of which the most common is usually chlorophyll-a which is considered to be the main centre of photosynthetic reactions (Buttery and Buzzell, 1977). Amount of chlorophyll-a in a waterbody may be estimated colorimetrically through cold acetone extraction.

Principle

In case most of the chlorophyll content of water is present as chlorophyll-a, an approximate estimate of chlorophyll may be obtained from the absorbance at 665 nm of a methanol or acetone extract. The results are expressed, as if all the chlorophyll were chlorophyll-a, as chlorophyll-a equivalent (Golterman et. al, 1978). The author has made some modifications of the commonly used method of chlorophyll-a estimation to suit the less sophisticated laboratories and has found the following procedure to perform fairly satisfactorily.

Requisites

(a) Apparatus etc: Colorimeter or spectrophotometer, centrifuge, plankton net.

(b) Glasswares : 50 ml volumetric flasks, centrifuge tubes, pipette.

(c) Reagents : Acetone. Available in market. This reagent should be kept in refrigerator.

Procedure

Filter a suitable volume of water through plankton net so that the filtrate contains approximately 0.01 - 0.05 mg chlorophyll -a at 50 ml volume and collect the plankton volume in a centrifuge tube. Centrifuge the volume so that the plankton are settled at the bottom of the tube. Decant the supernatent water. Rinse the plankton in the tube repeatedly with small volumes of cold acetone and transfer to a 50 ml volumetric flask. After removal of all the biomass to the flask, wash the tube a few times with acetone and transfer the solvent to the volumetric flask, make up the volume. Stopper the flask and keep in a refrigerator.

Bring out the flask after 24 hrs and measure the absorbance of the solution with the help of a colorimeter or spectrophotometer at 665 and 750 nm. respectively.

Calculation

If the observed absorbance at 665 nm and 750 nm are K_1 and K_2 and A = volume of acetone extract (50 ml here), L =length of light path in the spectrophotometer in cm and V =volume (ml) of filtered water sample, then concentration of chlorophyll-a equivalent in original water sample will be

$$11.9\ (K_1 - K_2) \times \frac{A}{L} \times \frac{1000}{V}\ m_g \ell^{-1}$$ (Vollenweider, 1969).

Levels of chlorophyll–a in pond water

Limited studies have been done on optimum or critical levels of chlorophyll-a content of fish pond water. However a range of 50 - 200 $\mu g\ \ell^{-1}$ is usually considered to be index of good productivity.

3

SOIL ANALYSIS

Bottom soils play a significant role in determining productivity of fish ponds. Occurrences of various nutrient elements in pond water and maintenance of different chemical conditions depend largely on the bottom soil of the fish pond wherein a series of chemical and biochemical reactions continuously take place resulting in release of different nutrient elements into the overlying water and their adsorption by the soil mass and the microbial population. These processes influence greatly the growth and abundance of different fish food organisms and, therefore, an intimate knowledge of the nature and properties of the pond soils is essential for increasing the productivity of fish ponds (Mandal and Chattopadhyay, 1992). Considering these importance of soils in maintaining productivity of fish culture systems, Hickling (1971) described bottom soils to be the chemical laboratory of fish ponds. In the present section the methods of determination of some important properties of fish pond soils have been discussed.

3.1 Collection of soil samples

As in case of water, collection of soil samples for different analyses needs to be attended carefully since any error at the time of sampling can not be corrected at later stages.

3.1–1 Procedures of soil sampling

The collected soil should be representative of the pond bottom. A pond can be treated as a single sampling unit only if the bottom of it exhibits appreciable uniformity with respect to depth, texture, organic matter content etc. In case there are considerable differences in such properties, separate samples need to be collected from such zones. For each sampling unit it is necessary to collect a number of sub-samples from different points scattered thoroughly over the area under study. About 1 kg of sub-sample is usually required to be collected covering about 6" depth at the surface of the bottom soil at each spot. This is considered to be the most reactive zone of bottom soil from the point of view of pond fertility. In some specific conditions, however, samples from greater depth of soil may be required. For example, in case of acid sulphate soils, the acidity is usually contributed from deeper horizon of the soils and for assessing the nature and properties of fish pond soils in acid sulphate zones, it may be necessary to collect soil samples from deeper horizons. After collection, put all the sub-samples on a clean polythene sheet or enamel tray and mix them thoroughly. Divide the total volume of soils into four parts, remove two of them and mix thoroughly the remaining two portions. Discard some amount of this sample and retain about one kg for subsequent analyses.

3.1-2 Processing of soil samples

Collected soil samples should be air dried in shade. Avoid drying the samples under high temperature or directly under the sun since this may bring about major changes in soil properties, especially mineralisation of soil organic matter. After drying, grind the soils with a porcelin or wooden pestle and morter, pass primarily through a 2 mm and then through a 80 mesh sieve. Store the soil sample in clean polythene container for the analysis.

Although such air dried soil samples are widely used for various analyses yet some soil properties like relative concentrations of NH_4^+ and NO_3^- nitrogen or availability of P may tend to change considerably on drying. If very accurate knowledges about various properties of pond soils are required, it is advised to analyse the soil samples as soon as possible after collection under field moist conditions. A separate portion of this soil sample may be taken for drying and the moisture content of the soil be determined from the differences. While interpreting the data, the result may be expressed in terms of dry weight of soils. In various standardised procedures, however, air dried pond soils have been widely used and such soil samples may be utilised for various common analyses.

3.2 Analysis of soils

There are numbers of soil factors which may influence the productivity of fish ponds. In the present section, analytical procedures of some common par meters have been described. Studies on all these properties are usually sufficient for assessment of productivity of general fish culture systems. However, for some specific reasons, there may be necessity of analysing some other parameters of the pond soils which have not been included in this section. In such cases readers may consult Piper (1967), Jackson (1971), Black (1965) or any other standard book on chemical analyses of soils. The author will be happy to satisfy any enquiry of the reader regarding such analyses.

3.2-1 pH

Role of pH in determining productivity of an aquaculture system has been discussed earlier (Sec. 2.2-1). Principles for estimation of pH of soils are basically the same viz. colorimetric and potentiometric. In case of soil samples, however, a soil-water suspension is used to determine the pH value.

A. Colorimetric method

This method is commonly used for field studies. However, as has been stated earlier, colorimetric determination of pH may not be very accurate.

Principle

Under colorimetric method of pH determination of soil samples, a soil-water suspension is prepared by using various soil: water ratios, more commonly 1 : 2. Barium sulphate ($BaSO_4$) is used to flocculate the suspension and to make the supernatent water clear. pH of this water is allowed to attain equilibrium with that of the soil and is determined with the help of a Lovibond Comparator as stated in sec. 2.2.-1.

Requisites

(a) Apparatus – Lovibond comparator

(b) Glasswares – 50 ml tubes with stoppers

(c) Reagents – (i) Indicator – As mentioned in sec. 2.2.-1.

(ii) Neutral $BaSO_4$ – AR grade reagent, available in the market. Dissolve 116.7 g $BaSO_4$ in 750 ml DW. Make up the volume to 1 l.

Procedure

Put some amount of $BaSO_4$ solution in a 50 ml tube to form about 1 cm thick layer. Add 10 g of soil sample and 20 ml of DW to the tube. Shake the contents thoroughly and keep for about half an hour to let the soil particles settle down. Then pipette out about 10 ml of supernatent water and determine its pH value colorimetrically as stated in sec. 2.2.-1.

B. Potentiometric method

Potentiometric method is much more reliable and has become extremely popular now a days. Portable battery operated pH meters are being widely used for field studies as well as in laboratories.

Principle

Principle of potentiometric determination of pH has been described in sec. 2.2.-1.

Requisites

(a) Apparatus – pH meter

(b) Glasswares – (i) 50 ml beakers (ii) glass rod (iii) measuring cylinder.

(c) Reagents – Buffer solutions - as stated in sec. 2.2.-1.

Procedure

Take 10 g of soil samples in a clean 50 ml beaker, add 20 ml DW, shake. Keep the soil-water suspension for at least half an hour with occasional stirring with a glass rod to establish an equilibrium. After the stipulated period, find out the pH of the soil-water suspension by following the method stated in sec. 2.2.-1.

Comments

Different aspects of colorimetric as well as potentiometric methods of pH estimation have been discussed in sec. 2.2.-1. Some additional information may be provided here with respect to determination of soil pH.

Soil pH values generally tend to change with variations in soil-water ratio and hence for getting comparable results uniform soil-water ratio should be maintained for different soil samples. The ratio should also be mentioned, wherever necessary, to denote the pH values. Some soils exhibit extreme variability in pH values when exposed to dry condition. In such cases, it is advisable to determine the pH of moist soil sample as soon as possible after collection. In such cases, an approximation of actual weights of the soil and water may be made and the volume of DW to be added may be adjusted accordingly.

pH levels in pond soils

As in case of water, importance of neutral to slightly alkaline pH range of fish pond soils has been emphasized by many workers for achieving good productivity. In India, Banerjea (1967) studied the pH values of large numbers of pond soils with relation to productions in those ponds. As per his observations neutral to slightly alkaline pH ranges of pond soils may be considered to be conducive to good aquaculture.

3.2-2 Lime Requirement

Production of fish is usually of low order in ponds with acidic soils. This problem of poor productivity may be solved by application of lime. Lime requirement (LR) denotes the amount of lime required to be added in a pond to bring the pH of the bottom soil to a desired level. Various levels of alkalinity required to be raised by lime application have been suggested by many workers. However, probably the most common opinion is to bring the soil pH to near neutral. Boyd (1974), on the other hand, suggested that lime should be applied to raise the total hardness of a pond water to 20 ppm.

Two methods of LR determination are commonly followed in most of the laboratories. First one is more simple to perform where

LR is determined from textural composition and pH of the soils while in the second method, a buffer solution is used for this purpose. Both of these two methods have been discussed in this section.

A. Determination of lime requirement from data on texture and pH.

Principle

Requirement of lime in a pond is controlled mainly by two factors pH and cation exchange capacity of soils of which the second property of the soil depends largely on its textural composition. Many workers (Schaeperclaus, 1933 and others) have used these two properties for assessing LR of pond soils. The suggestion from CIFRI (1985) for Indian soils has been dealt here.

Requisites

Same as in case of determination of pH and textural composition (sec. 3.2-1 and 3.2-13).

Procedure

Determine pH and textural composition of the soil by following the methods described in sec. 3.2-1 and 3.2-13.

Calculation

Use Table 7 (CIFRI, 1985) for determining LR of the pond soil as kg ha^{-1} of $CaCO_3$.

Table 7: Rates of lime ($CaCO_3$) for application in ponds

Soil pH	*Dose of lime (kg ha^{-1})*		
	Sandy soil	*Med. soil*	*Clayey soil*
4.0 – 5.0	1000	2000	3000
5.0 – 6.0	600	1200	1800
6.0 – 6.5	500	1000	1500
* 6.5 – 7.5	200	400	600

* Though this group does not fall under acidic category, yet liming is advised as a prophylactic measure.

B. Determination of lime requirement by using buffer solutions.

Principle

This method, as described by Boyd (1974), deals with determination of pH of pond soils through use of a buffer solution of pH 8.0 and assessing the LR of the soils for raising the total hardness to 20 ppm or more.

Requisites

(a) Apparatus : pH meter

(b) Glasswares : 100 ml beakes, 1ℓ vol. flask, pipette etc.

(c) Reagents : (i) Buffer solution : Dissolve 20 g p-nitrophenol, 15 g boric acid, 74 g potassium chloride and 10.5 g potassium hydroxide in about 750 ml DW. Make up the volume to 1ℓ. This buffer solution gives a pH value of 8.0 ± 0.1

(ii) Other reagents : Buffer solutions used for pH estimation, as stated in sec. 3.2-1

Procedure

Take 20 g of soil sample in a 100 ml beaker, add 20 ml DW and stir intermittently for half an hour. Measure pH of the suspension as described in sec. 3.2-1. Then add 20 ml of buffer solution(pH 8.0) to this soil : water suspension. Keep for about half an hour with occasional stirring. Set the pH meter reading to 8.0 with 1 : 1 mixture of buffer solution and DW. Under this standardisation, note the pH reading of the soil : water + buffer suspension from the pH meter.

Calculation

Find out the LR of the soil from Table 8, prepared by slight modification of Boyd (1974).

Table 8 : L.R. in kg ha^{-1} of $CaCO_3$ for fish pond soils.

Soil pH	*pH of soil : water + buffer*								
	7.9	*7.8*	*7.7*	*7.6*	*7.5*	*7.4*	*7.3*	*7.2*	*7.1*
	Requirement of $CaCO_3$								
5.7	90	180	270	360	450	540	630	725	820
5.6	125	250	380	500	630	700	880	1010	1130
5.5	200	400	600	810	1010	1210	1410	1610	1810
5.4	290	580	870	1160	1450	1740	2030	2320	2610
5.3	340	680	1020	1360	1700	2040	2380	2720	3060
5.1	440	880	1325	1760	2200	2650	3090	3530	3970
4.9	660	1310	1970	2620	3280	3930	4590	5240	5980
4.8	670	1340	2020	2690	3360	4030	4700	5390	6050
4.7	700	1410	2120	2820	3530	4230	4940	5640	6350

Comments

Besides limestone ($CaCO_3$), quick lime (CaO) or slaked lime ($Ca(OH)_2$) are also used as liming materials. When applied in equivalent amount, all these liming materials should produce similar results. However, the reaction will be faster with CaO and $Ca(OH)_2$. On the other hand, $CaCO_3$ is considered to be the safest, cheapest and most widely used liming material in fish culture.

3.2–3 Gypsum requirement

In alkali soils Na usually occurs in high concentrations to result in very high pH values alongwith associated problems and needs to be recovered from the exchange complex by divalent cations, especially Ca. Gypsum ($CaSO_4$) is widely used for replacing the exchangeable Na^+ and the amount of gypsum needed to substitute the exchangeable Na^+ to a safe level is known as gypsum requirement (GR) of soils. GR is not usually included in fish pond soil analysis programmes but this may become important in case of pond soils exhibiting very high pH values.

Principle

A simple method for determination of GR has been described by USDA (1968). GR of a soil can be conveniently determined by allowing a known concentration of $CaSO_4$ solution to react with the soil. The amount of divalent cations (Ca^{++} + Mg^{++}) in the solution is then estimated and the amount of gypsum reacted with the soil is worked out.

Requisites

(a) Glasswares : 250 ml bottles, all glasswares required for Ca + Mg estimation, as stated in sec. 2.2-8.

(b) Reagents :
(i) Standard gypsum solution : Add 5 g $CaSO_4$, $2H_2O$ to 1ℓ DW, shake intermittantly for about half an hour, filter and determine the Ca concentration in the filtrate by following sec. 2.2.-8.
(ii) Other reagents : Prepare all the reagents for Ca and Ca + Mg estimation by following sec. 2.2-8.

Procedure

Take 5 g soil sample in a 250 ml bottle and add 100 ml of standard $CaSO_4$ solution to it. Stopper the bottle and shake time to time for about half an hour. Then filter the suspension and determine the Ca + Mg concentration in the filtrate by following EDTA titration method using eriochrome black T indicator as described in sec. 2.2.-8

Calculation

(i) Meq ℓ^{-1} of Ca + Mg in the filtrate = (mℓ of EDTA used X normality of EDTA solution × 1000)/ml of aliquot used.

(ii) GR (meq $100g^{-1}$) = Conc. of Ca in added $CaSO_4$ solution (meq ℓ^{-1}) – Conc. of Ca + Mg in filtrate (meq ℓ^{-1}) × 2.

3.2–4 Alkaline earth carbonate

Alkaline earth carbonate is also referred as free calcium carbonate ($CaCO_3$) content of soils and provides an idea about lime content and hence the alkalinity of soils.

Principle

Alkaline earth carbonate content of a soil can be determined rapidly by solubilising the $CaCO_3$ in standard hydrochloric acid (HCl) and estimating the amount of HCl consumed for this purpose.

Requisites

(a) Apparatus : Mechanical shaker.

(b) Glasswares : 250 ml bottles, 100 ml cylinder, 100 ml conical flask, bottle, pipette etc.

(c) Reagents :

(i) 0.5 N HCl : Dilute 50 ml concentrate HCl to 1ℓ. Standardise the strength of the solution to 0.5 N by titration with 0.5 N Na_2CO_3 (53.0 g Na_2CO_3 dissolved in 1 ℓ DW) using bromothymol blue indicator.

(ii) 0.5 N NaOH : Dissolve 20 g NaOH in 1ℓ DW. Standardise the strength of the solution to exactly 0.5 N by titrating with standard 0.5 N HCl using bromothymol blue indicator.

(iii) Bromothymol blue indicator : Dissolve 0.5 g of indicator in 100 ml absolute alcohol.

Procedure

Take 5 g of soil sample in a 250 ml bottle, add 100 ml of 0.5 N HCl and shake for 1 hr. Allow the soil to settle down. Then pipette out 20 ml of supernatent liquid and take in a 100 ml conical flask. Add a few drops of bromothymol blue indicator to this liquid to develop a yellow colour. Titrate it with 0.5 N NaOH till the colour turns blue. Carry out a blank set without any soil sample but using all the reagents in the same way.

Calculation

Alkaline earth carbonate ($CaCO_3$ equivalent)% = Blank titration value (ml) – soil solution titration value (ml) × 2.5.

Comments

The value of $CaCO_3$ equivalent obtained through this method may be slightly higher than the actual value because soil components other than lime may also react with the acid. However, this value can be safely used for most of the purposes.

3.2–5 Electrical Conductivity

Electrical conductivity (EC) indicates the total concentration of ionised constituents of a system. It is closely related to the sum of cations or anions, as determined chemically, and usually correlates with the amount of total soluble solids. Much information on effect of EC levels on productivity of ponds is not available. However, considering that changes in EC are associated with release or depletion of soluble ions in soil-water system, EC might have an indirect role to play in pond productivity (Mandal and Chattopadhyay, 1990). EC values can be used to measure the extent of soil salinity also and may be of special significance for brackishwater fish ponds.

Principle

EC can be measured with the help of a conductivity bridge, principle and details of which have been discussed in sec. 2.2.-2. For determination of EC of soils, a saturated soil paste should be preferably used. However, preparation of saturation extract requires some specialisation and is also a bit time consuming. Hence, for most practical purposes, 1 : 2 soil : water suspension is used.

Requisites

(a) Apparatus : Conductivity meter

(b) Glasswares : 100 ml beakers, glass rod, measuring cylinder etc.

(c) Reagents : (i) 0.02 M Potassium Chloride (KCl) : Prepare the reagent as stated in section 2.2.-2.

Procedure

The soil : water suspension prepared for pH estimation may be used for EC estimation as well or a fresh soil : water suspension (1 : 2) may be prepared as stated in sec. 3.2-1. Keep the soil water suspension for at least half an hour with intermittent stirring. Measure the cell constant and then the EC of the suspension from the conductivity meter by following the method described in sec. 2.2-2.

Calculation

Calculate the EC of the soil from the relationship, EC (mmhos cm^{-1}) = Measured conductivity (mmhos cm^{-1}) × Cell constant

Comments

Effect of temperature on EC values has been discussed under sec. 2.2-2. Regarding use of soil : water suspension it may be worthwhile to mention that EC values decline sharply with increase in the amount of water in the soil : water suspension depending on the texture and composition of salts (Jackson, 1971). In this book, 1 : 2 soil : water ratio has been used for determination of EC of soils. However, other ratios like 1 : 1, 1 : 2 5 or 1 : 5 are also used. For any E.C value, therefore, the soil : water ratio should be mentioned clearly (e.g. $EC_{1:2}$, $EC_{1:1}$ etc).

3.2-6 Organic Carbon

The role of organic carbon (OC) in maintaining the productivity of fish ponds is well recognised. Apart from influencing various physico-chemical properties of bottom soils and releasing different nutrient elements to more available forms in the pond environment, it controls an important property of the pond eco-system viz. oxidation -reduction reaction (Mandal and Chattopadhyay, 1992).

Principle

The rapid titrimetric method of Walkley and Black (1934) using heat of dilution is widely used for determination of OC in soils. This method has an advantage that it excludes less active elementary carbon of soil (e.g. graphite) and includes only that part of OC which is involved in maintenance of pond productivity. Under this method organic matter in soil is oxidised with excess standard potassium dichromate ($K_2 Cr_2O_7$) using the heat of dilution of added concentrated H_2SO_4. The unutilised $K_2Cr_2O_7$ is then titrated with standard ferrous ammonium sulphate ($Fe(NH_4)_2(SO_4)_2$) and the amount of OC is determined from the amount of standard $K_2Cr_2O_7$ used for oxidation.

Requisites

(a) Glasswares : 500 ml conical flasks, pipette, burette, measuring cylinder, 1 ℓ volumetric flasks.

(b) Reagents

  (i) 1 N $K_2Cr_2O_7$: Dissolve 49.0 g of solid $K_2Cr_2O_7$ in DW and make up the volume to 1 ℓ.

  (ii) 1N $Fe(NH_4)_2(SO_4)_2$: Dissolve 392.2 g of $Fe(NH_4)_2(SO_4)_2$, $6H_2O$ in 800 ml of DW, add 20 ml conc. H_2SO_4 and dilute to 1 ℓ.

  (iii) Diphenyl amine indicator : Dissolve 0.5 g diphenyl amine in 20 ml DW and add 100 ml conc. $H_2 SO_4$.

  (iv) Conc. H_2SO_4.

  (v) 85 per cent orthophosphoric acid ($H_3 PO_4$).

Procedure

Take 1g of soil sample in a 500 ml conical flask and moisten it with a few drops of DW, keep for about 10 minutes. Add exactly 10 ml of 1 N $K_2Cr_2O_7$ and then 20 ml of conc. H_2SO_4. Mix the contents thoroughly and keep in a dark place for 30 minutes. Then add 200 ml DW and 10 ml of H_3PO_4. Put about 1 ml of diphenyl amine indicator to develop a deep blue colour. Titrate the solution with standard $Fe(NH_4)_2(SO_4)_2$. At the end point the blue colour suddenly flushes to bright green. Carry out a blank with all the reagents but without the soil.

Calculation

Organic Carbon (per cent) = $(B - A) \times 0.3$

Where, B = titration value (ml) of $Fe(NH_4)_2(SO_4)_2$ in blank set.

A = titration value (ml) of $Fe(NH_4)_2(SO_4)_2$ with soil.

Comments

Chloride ions, when present in considerable concentrations in soil, may interfere with O.C determination under this procedure. Such interference can be controlled by application of silver sulfate ($AgSO_4$). Add 5 g of $AgSO_4$ in 100 ml conc. H_2SO_4. Use 20 ml of this $AgSO_4$ mixed H_2SO_4 after addition of $K_2Cr_2O_7$ for digestion in stead of ordinary H_2SO_4. Such modification is essential for estimating OC content of brackishwater pond soils.

OC levels in pond soils

Banerjea (1967) categorised fish pond soils of India under different productivity levels with respect to OC status of the soils. In his opinion, pond soils having less than 0.5 per cent OC may be considered to be too poor while those in the range of 1.5 – 2.5 per cent may be taken as optimum for fish production.

3.2–7 Cation Exchange Capacity

The total quantity of cations which a soil can adsorb by cation exchange is termed as cation exchange capacity (CEC) of the soil and is expressed as milliequivalents per 100 g soil (USDA, 1961). Since amount of easily exchangeable cations increase under soils with higher CEC values, such soils are likely to provide more amount of positively charged nutrient elements in easily available form. Close relationship between CEC of bottom soils and productivity of ponds have been reported by Anon (1982).

Principle

For determination of CEC the soil exchange phase is first saturated with sodium (Na^+) ions. These Na^+ ions are again replaced by any other suitable cation and the amount of Na^+ ions released from the exchange site are measured as index of CEC.

Requisites

(a) Apparatus: Mechanical shaker, flame photometer, centrifuge.

(b) Glasswares: 250 ml narrow mouth glass bottles, 250 ml conical flasks, funnels, pipettes etc.

(c) Reagents :

(i) 1 N Sodium acetate (CH_3COONa) solution: Dissolve 136.0 g sodium acetate trihydrate ($CH_3COONa, 3H_2O$) in about 750 ml DW and make up the volume to 1 ℓ. Adjust the pH of the solution to 8.2 with dilute NaOH.

(ii) Ethyl alcohol (95 per cent).

(iii) 1 N Ammonium acetate solution (CH_3COONH_4) : Add 57 ml conc. acetic acid (CH_3COOH) to about 750 ml DW. Follow with addition of 68 ml conc. ammonium hydroxide (NH_4OH). Make the total volume to 1 ℓ. Adjust the pH of the solution to 8.0 by addition of dil. NH_4OH or CH_3COOH, as required.

Procedure

Take 5 g of soil sample in a 250 ml bottle and add 100 ml of 1 N CH_3COONa solution, shake for half an hour. Centrifuge the contents at 2400 rpm to precipitate the soil at the bottom and decant the supernatent solution. In case centrifuge is not available in the laboratory, keep the bottle horizontally overnight and on the next day decant the supernatent solution very carefully. Add about 35 ml of ethyl alcohol to the bottle, shake well, either centrifuge and decant the supernatent liquid or keep it for some time and filter the supernatent solution carefully without disturbing the soil as far as possible. Repeat this procedure three times and if filtration is done, use the same filter paper for all the filtrations of the same sample. Put the filter paper inside the bottle with soil sample and add 100 ml of 1 N CH_3COONH_4 solution to the bottle and shake for half an hour. Filter the total contents of the bottle in a clean 1 ℓ volumetric flask. Make up the volume of the filtrate to 1 ℓ with DW and determine the concentration of Na^+ ions in this solution flame photometrically, as discussed in sec. 2.2.-8.

Calculation

If concentration of Na^+ ions in extracted solution is x ppm, then,

CEC of the soil sample will be $\frac{20x}{23}$ meq per 100 g soil.

Comments

In recent years CEC of soil is being widely expressed as c mol (P^+) kg^{-1}. However, in this book the old and more acquainted unit (meq per 100 g soil) has been used.

3.2-8 Exchangeable bases (calcium, magnesium, sodium, potassium)

Different exchangeable cations differ in their effects on acidity or alkalinity of soils. While H^+ and Al^{+++} ions tend to increase the acidity of soils, most of the other cations neutralise soil acidity and increase alkalinity. These cations are called exchangeable bases and are usually dominated by Ca^{++}, Mg^{++}, Na^+ and K^+ ions.

Principle

Exchangeable Ca^{++}, Mg^{++}, Na^+ and K^+ can be estimated by using standard procedures after replacing these cations from the exchange complex by NH_4^+ ions using neutral normal ammonium acetate (CH_3COONH_4). Respective bases can then be determined by using the procedures described in sec. 2.2.5, 2.2.8 and 2.2.7.

Requisites

(a) Apparatus : Mechanical shaker

(b) Glasswares : 250 ml conical flasks, 1 ℓ vol. flasks, pipette, burette, 100 ml cylinder etc.

(c) Reagents :

(i) Neutral normal ammonium acetate (CH_3COONH_4) solution : Add to 750 ml DW, 57 ml conc. acetic acid (CH_3COOH) and 68 ml conc. ammonium hydroxide (NH_4OH). Mix thoroughly and dilute the volume to 1 ℓ. Adjust the pH to 7.0 by gradual addition of NH_4OH or CH_3COOH.

(ii) 80 per cent Alcohol : Dilute 400 ml ethyl alcohol to 500 ml with DW.

(iii) Other reagents : Same as in case of determination of Ca, Mg, Na and K (secs. 2.2-5, 2.2-8, 2.2-7).

Procedure

Weigh 5 g of soil sample in a 250 ml bottle. Add 100 ml 80 per cent alcohol, shake for five minutes and wait for about 2 hrs so that the suspended colloids settle down and then filter the supernatent solution through a Whatman No. 42 filter paper. After filtration, open the wet filter paper and place it on a funnel. Put the funnel on the mouth of the bottle. Take 100 ml neutral, normal ammonium acetate in a 100 ml cylinder. Pipette out ammonium acetate from this cylinder and wash the filter paper in such a way that the residual soils in the filter paper are washed down to the soil sample in the bottle. Use entire 100 ml volume of ammonium acetate in this process. Stopper the bottle and shake for half an hour in a mechanical shaker. Filter the entire suspension through the same filter paper used earlier.

Take required amount of aliquots from this filtrate and determine Ca, Mg, Na and K concentration in this solution by following the methods described in secs. 2.2.-5, 2.2-8 and 2.2-7.

Calculations

Determine the concentration of respective cations in the extracted solution as ppm from the flame photometer reading.

(a) Concentration of exchangeable Ca^{++} (Meq per 100 g soil)

$$= \text{ppm of Ca in the extract} \times \frac{1}{10}$$

(b) Concentration of exchangeable Mg^{++} (Meq per 100 g soil)

$$= \text{ppm of Mg in the extract} \times \frac{1}{6}$$

(c) Concentration of exchangeable Na^{+} (Meq per 100 g soil)

$$= \text{ppm of Na in the extract} \times \frac{1}{11.5}$$

(d) Concentration of exchangeable K^{+} (Meq per 100 g soil)

$$= \text{ppm of K in the extract} \times \frac{1}{19.5}$$

Comments

As stated in the previous section, now a days c mol (P^+) Kg^{-1} is widely being used to express amount of exchangeable ions in soils, especially agricultural ones. But this system is not so much in vogue in case of fish pond soils. In this communication, therefore, the old unit has been used.

3.2-9 Exchangeable sodium percentage

It is often convenient to express the occurrence of an exchangeable cation with relation to total exchange capacity instead of expressing the concentration of the particular cation in isolation. This expression clearly indicates the dominance of the cation under study over the others. The degree of saturation of the soil exchange complex with sodium ions is termed as exchangeable sodium percentage (ESP).

Higher ESP values indicate occurrence of relatively higher concentrations of Na^+ ion in the exchange complex and the effects of Na^+ ions will be more pronounced in such soil although actual concentration of Na^+ ion in this soil may be lower than some other soil with high concentration of Na^+ ion but lower ESP value.

Principle

ESP of a soil can be determined from its exchangeable Na^+ and cation exchange capacity value using the following formulae:

$$\text{ESP} = \frac{\text{Exch. Na}^+(\text{meq per 100 g soil})}{\text{CEC (meq per 100 g soil)}} \times 100$$

Requisites

All the materials for determinations of CEC (sec. 3.2-7) and exchangeable Na^+ (sec. 3.2-8).

Procedure

Determine the amount of exchangeable Na^+ and CEC value in the soil by following the methods described in sections 3.2-8 and 3.2-7). Use these two values to calculate ESP from the formulae stated above.

Comments

Na^+ ions may occur in high concentrations in fish ponds situated in alkaline areas. It may be useful to determine ESP in such soils.

3.2-10 Nitrogen (Easily mineralisable nitrogen, mineralised nitrogen and total nitrogen)

Nitrogen (N), being a major nutrient element, plays an important role in determining the fertility of fish ponds. Hence determination of available nitrogen in pond soils is an important prerequisite for assessing the fertility status of the culture system. In soils nitrogen (N) occurs mostly in organic form. From this state it is gradually mineralised to inorganic forms and becomes available to primary producers. Jhingran et. al. (1968) described determination of easily mineralisable form of N to be well suited for assessing availability of N in fish pond soils. However, the mineralised (NH_4^+ and NO_3^-) forms of N in pond soils are more readily available to phytoplankton population than easily mineralisable forms and thus is likely to serve as a better index of immediately available N. Chattopadhyay et. al. (1985) also observed mineralised ($NH_4^+ + NO_3^-$) forms of N in soils to exhibit better correlation with production of primary fish food organisms in brackishwater fish ponds. In the present section, therefore, the methods of determination of easily mineralisable and mineralised forms of N and also total N have been discussed.

A. Easily mineralisable nitrogen

Fractions of organic nitrogen like amino acid, peptides or easily decomposable proteins constitute what is known as easily mineralisable N. The amount of N under these states are easily transformed into readily available forms and may thus provide an index of available nitrogen status of the pond soils.

Principle

The amount of N under easily mineralisable form can be determined by oxidising the soil organic matter with a mild oxidising agent so that only the easily mineralisable forms of organic nitrogen are oxidised. 0.32 per cent potassium permanganate

($KMnO_4$) is commonly used for this purpose (Subbiah and Asija, 1956). The released NH_3 is absorbed in H_3BO_3 as stated in sec. 2.2-12

Requisites

(a) Apparatus : Kjeldahl distillation set

(b) Glass wares : Kjeldahl flask (800 ml), 150 ml conical flask, pipette, burette etc.

(c) Reagents

(i) 0.32 per cent $KMnO_4$: Dissolve 3.2 g $KMnO_4$ in 1 ℓ DW.

(ii) 2.5 per cent NaOH : Dissolve 25 g NaOH in about 800 ml DW, cool and make the volume to 1 ℓ.

(iii) 0.02 N H_2SO_4 : Prepare the reagent by following the method described in Appendix -1

(iv) H_3BO_3 indicator solution : Prepare by following the method described in sec. 2.2-12

(v) Liquid paraffin (pure) : Available in market.

Procedure

Take 20 g of soil sample in a Kjeldahl flask, add 20 ml DW and mix thoroughly by swirling the flask. Add about 1 ml of liquid paraffin and a few glass beads in the flask to prevent bumping and frothing. Add 100 ml of 0.32 per cent $KMnO_4$ and 100 ml of 2.5 per cent NaOH to the flask. Connect the condenser mouth immediately and distill the contents to collect nearly 100 ml of the distillate in H_3BO_3 solution (20ml) in a receiving conical flask. Titrate the distillate with 0.02 N H_2SO_4 to development of a pink colour.

Calculations

If the volume of 0.02 N H_2SO_4 required for titration is x ml,then amount of easily mineralisable N in soil will be x X 1.4 mg per 100 g soil.

B. Mineralised ($NH_4^+ + NO_3^-$) nitrogen

After mineralisation, organic N is transformed into two major inorganic forms viz. ammonium (NH_4^+) and nitrate (NO_3^-). Of these two forms, which are readily available to phytoplankton,

NH_4^+ ions largely remain adsorbed in the exchange complex of soil colloids as easily available exchangeable cations. NO_3^- ions, on the other hand, become available mostly in water phase.

Principles

N in NH_4^+ form can be determined by extracting these ions with acidified potassium chloride (KCl) solution, as stated by Jackson (1971). This extract also contains the NO_3^- form of N which is readily soluble and total amount of N under both of these two forms can be determined by using Davarda's alloy, as stated in sec. 2.2-12.

Requisites

(a) Apparatus—Kjeldahl distillation set.

(b) Glasswares: Kjeldahl flask (800 ml), 150 and 500 ml conical flasks, 1 ℓ volumetric flask, pipettes, burette etc.

Reagents

(i) 10 per cent Sodium chloride (pH 2.5):

Dissolve 100 g sodium chloride (NaCl) in about 100 ml DW. Add a few drops of 2-4 dinitrophenol indicator to get yellow colour. Now add dilute (1 : 10) HCI continuously till the yellow colour disappears. Add 5-6 drops of HCl more and make up the volume to 1 ℓ.

(ii) 2-4 Dinitrophenol indicator: Prepare by following the method described in sec. 2.2-13.

(iii) 0.02 N H_2SO_4 : Prepare by following the method described in appendix 1.

(iv) Boric acid (H_3BO_3) - Mixed indicator: Prepare by following the method described in sec. 2.2-12

(v) 40 per cent Sodium hydroxide (NaOH): Add 40 g NaOH in 100 ml DW.

(vi) Davarda's alloy: Available in market.

Procedure

Take 100 g soil in a 500 ml conical flask and add 200 ml NaCl (pH 2.5). Shake the contents thoroughly at first and then intermittantly for half an hour. Filter the suspension and wash the soil with additional 50 ml of NaCl. Take the filtrate in a Kjeldahl flask, add about 1 ml of liquid paraffin, a few glass beads and 0.5 g of Davarda's alloy. Add 20 ml 40 per cent NaOH and distill as stated in sec. 2.2-12 to collect the distillate in 20 ml H_3BO_3. Continue the distillation to near dryness and titrate the H_3BO_3 with standard (0.02 N) H_2SO_4 to development of a pink colour (sec. 2.2-12).

Calculations :

If the vol. of 0.02 N H_2SO_4 required for titration is x ml then the amount of mineralised ($NH_4^+ + NO_3^-$) N in the soil will be x X 0.28 mg N per 100 g soil.

C. Total nitrogen

Most of the nitrogen (N) in soil occurs in organic form from which state this nutrient element is gradually released to available form. Although not related with productivity of ponds directly, determination of total N presents an assessment of the N supplying capacity of a soil and forms an important parameter involved with productivity of fish ponds.

Principle

Two analytical procedures are widely used for determining total N content of soils—the Kjeldahl conversion of N to ammonium sulphate and Duma's conversion of nitrogen gas (Jackson, 1971). The former method is much more widely used and has been described here.

Requisites

(a) Apparatus : Kjeldahl digestion and distillation set.

(b) Glasswares : Kjeldahl flasks (800 ml), conical flasks (150 ml), pipettes, burette etc.

(c) Reagents

  (i) 40 per cent NaOH : Prepare this reagent by following the method described in part B of this section.

  (ii) Boric acid- indicator : Prepare this reagent by following section 2.2-12.

(iii) Digestion mixture : Mix approximately 20 g copper sulphate ($CuSO_4$, $5H_2O$), 3 g mercuric oxide (HgO) and 1g selenium (Se) powder. Add this mixture to 475 g sodium sulphate (Na_2SO_4) and mix thoroughly. Potassium sulphate (K_2SO_4) may also be used in place of Na_2SO_4.

CuSO$_4$, HgO and Se act as catalyst of the reaction while the purpose of adding Na_2SO_4 is to control the boiling point of the H_2SO_4 digestion (Jackson, 1971).

(iv) H_2SO_4 mixture : Add 3 g salicylic acid to 100 ml conc. H_2SO_4.

(v) Liquid paraffin : Available in market.

(vi) Glass beads ; Available in market.

(vii) 0.1N H_2SO_4 : Prepare this reagent by following appendix.

Procedure

Take 5 g soil sample in the Kjeldahl flask, add 20 ml DW and keep the sample for about half an hour. Then add slowly 35 ml of conc. H_2SO_4 mixture. Rotate the flask gently so that the acid mixes thoroughly with the soil part. Add about 10 g of the digestion mixture, shake gently to ensure effective mixing. Start heating the sample at 100°C temperature for about 20 minutes and then carry out the digestion at temperature between 360- 410°C. Continue boiling till the colour of the mixture becomes more or less transparent. Generally it takes abut one and half an hour to complete digestion. Cool the mixture after digestion and add about 250 ml DW to it.

Fit the Kjeldahl flask containing cooled, digested sample to a distillation set. Add a few glass beads and about 1 ml of liquid paraffin. Take 25 ml 4 percent H_3BO_3–indicator in an 150 ml conical flask and fit the flask at the receiving end of the the distillation set so that the delivery tube of the distillation set dips into the boric acid. Check all the fittings of the arrangement to avoid any possibility of loss of N.

Add slowly about 100 ml 40 per cent NaOH to the solution and close the mouth of the flask immediately with connector of the distillation set. Shake the flask slowly without disturbing the fittings. Then start heating, initially at a low temperature and then increasing the heat gradually. Collect about 150 ml distillate in H_3BO_3 in the receiving conical flask, stop heating and remove the flask from the apparatus to avoid sucking back.

Titrate the H_3BO_3 with standard 0.1 N H_2SO_4, as stated previously (sec. 2.2-12).

Calculations

If the volume of 0.1N H_2SO_4 required for titration of the sample be x,

$$\text{concentration of N in soil(\%)} = \frac{x \times 1.4}{W}$$

Where W = Wt of the soil (g) taken

Comments

Sucking back of the distillate into the sample solution may sometime occur due to heating at low temperature. In such case, the heating should be stopped, the entire remaining amount of the distillate be transferred to the kjeldahl flask again and a fresh distillation should be carried out with the same sample using fresh volume of H_3BO_3.

Levels of N in pond soils

Banerjea (1967) studied the easily mineralisable nitrogen content of large numbers of freshwater fish pond soils and suggested that the amount below 25 mg N per 100 g soil may be considered to be indicator of low productivity while the range of 25 to 75 mg N per 100 g soil may be considered as index of moderate to high nitrogen supplying capacity.

Very little work has so far been done on specific impact of total N status of pond soils on aquatic productivity. However, considering the relationship of total N with easily mineralisable forms and also organic C status of soils it may be approximated that soils with less than 0.05 per cent total N will indicate poor productivity while more than 0.1 per cent total N may be index of good productivity.

3.2-11 Phosphorus (Available phosphorus, total phosphorus, various inorganic forms)

Although phosphorus (P) usually occurs in very small amount in fish pond soil and water, this element is generally considered to be the most critical single factor in the maintenance of pond fertility (Jhingran, 1975). It is the available form of P in the soil and water phase which is important for aquatic productivity because phosphate ions in fish pond soils form insoluble compounds with iron and aluminium under acidic conditions and with calcium under alkaline environment rendering most of the P into unavailable forms in spite of the possibility of occurrence of this element in quite large amount in the soil (Banerjea and Mandal, 1965). Hence, for understanding the chemistry of P transformation in fish ponds, it may be necessary to know the occurrence of total amount of P and also different inorganic forms of the element, in addition to its concentration in available form. Present section deals with determinations of all these forms of P in pond soils.

A. Available phosphorus

Different methods are used for determining the amount of available phosphorus (P) in soils and the choice of a suitable method depends largely on the nature and properties of the soils (Benerjee and Chattopadhyay, 1980). Availability of P in a soil is contributed largely by various inorganic fractions of soil P. Magnitude of distribution of these forms depends largely on some particular soil properties, especially the pH of the soils. Jhingran *et. al* (1968) suggested that Truog's method may be used to assess the availability of P in fish ponds. In brackishwater fish ponds the soils usually exhibit alkaline reaction and Chattopadhyay et. al (1991) observed Olsen's method of determination of available P to be more closely correlated with production of benthic algae in such pond soils. Jackson (1971) also stated that in case of agricultural crops a higher correlation of soil test with crop response is obtained with Olsen's method of available P determination when all soils are considered, including acid, neutral, alkaline and calcareous. Both the methods have been discussed in this section.

A.1. Determination of available P by Truog's method

A number of dilute acid extraction methods have been proposed for measuring availability of P in soils. In general, the P extractable from acid soils by dilute solutions of strong acids can be correlated with crop yield response to P fertilisers. The correlation is lower with neutral soils but little or no correlation is obtained for soils that are alkaline (Jackson, 1971). However, although most of the freshwater pond soils exhibit neutral to alkaline reactions, yet Truog's method of extraction of soil P with 0.002 N H_2SO_4 has remained a widely used method of available P estimation in such pond soils.

Principle

Determination of available P by Truog's method involves shaking the soil sample with 0.002 N H_2SO_4, buffered at pH 3.0, and determination of P concentration in the extract.

Requisites

(A) Apparatus: (i) Colorimeter or Spectrophotometer
(ii) Mechanical shaker.

(b) Glasswares: (i) 250 ml bottles, (ii) 250 ml conical flasks, (iii) Funnels , (iv) 1 ℓ, volumetric flask (v) 25 ml volumetric flasks, (vi) Pipettes, (vii) Burrette etc.

(c) Reagents :

(i) 0.002 N H_2SO_4 : A stock solution of 0.1 N H_2SO_4 is prepared by following appendix. 20 ml of this acid is diluted to 1 ℓ to get 0.002 N H_2SO_4. Add 3 g of ammonium sulfate $(NH_4)_2SO_4$ to this volume before final volume of 1 ℓ is made.

(ii) Ammonium molybdate ($(NH_4)_6$ Mo_7O_{24}, $4H_2O$)- Potassium antimony tartarate solution : Prepare the reagent by following sec. 2.2-13

(iii) Ascorbic acid- molybdate solution : Use the method described in sec. 2.2-13

(iv) Standard (2 ppm) P solution : Prepare by following the method stated in sec. 2.2-13.

Procedure

Take 1.0 g soil sample in a 250 ml bottle and add 200 ml of 0.002 N H_2SO_4 (pH 3.0). Shake the suspension for half an hour in a mechanical shaker and then filter the suspension to estimate the concentration of P by following the method described in sec. 2.2-13.

Calculation

If the colorimetric reading for 10 ml aliquot corresponds to x ppm of P in the solution of the volumetric flask (25 ml), the amount of extractable (available) P in the soil will be x x 500ppm.

A. 2. Determination of available P by Olsen's method

Olsen's method is widely used for determining availability of P in alkaline or calcareous soils and has been observed to correlate closely with production of benthic algae in brackishwater fish ponds (Chattopadhyay et. al. 1981). Detailed studies have not yet been done with application of this method in case of freshwater pond soils. However, in view of the occurrence of neutral to alkaline pH ranges in productive freshwater ponds, Olsen's method may be used for determination of available P in freshwater pond soils as well.

Principle

This method (Olsen *et al*, 1954) involves extraction of soil P by a solution of 0.5 N sodium bicarbonate ($NaHCO_3$) adjusted to pH 8.5. The extraction procedure decreases the activity of Ca in soils and thus helps some P to be extracted from calcium phosphate form. In addition, this reagent also extracts some P from the surface of aluminium and iron phosphates (Jackson, 1971).

Requisites

(a) Apparatus : (i) Colorimeter or Spectrophotometer.

(ii) Mechanical shaker

(b) Glasswares : (i) 250 ml stoppered bottles, 1 ℓ, 100 ml, 25 ml volumetric flasks.

(c) Reagents :

(i) 0.5 M Sodium bicarbonate ($NaHCO_3$). Dissolve 42.0 g $NaHCO_3$ in 1 ℓ DW, adjust the pH to 8.5 through gradual addition of dilute NaOH using a pH paper. Put a thin layer of P free mineral oil on the solution if the solution is to be kept for longer period.

(ii) Activated charcoal : Available in market. Ordinary grade activated charcoal may contain some amount of P and to remove it the material may be leached with dilute $NaHCO_3$, dried and used in the study.

(iii) Ammonium molybdate — Potassium antimony tartarate solution : Prepare the reagent by following sec. 2.2-13.

(iv) Ascorbic acid — molybdate solution : Prepare by following sec. 2.2-13.

(v) 2—4 Dinitrophenol (DNP) indicator : Prepare by following the method described in sec. 2.8-3.

(vi) Dilute HCl : Prepare by diluting conc. HCl with DW using 1 : 10 ratio.

Procedure

Take 5 g soil sample in a 250 ml bottle, add 100 ml of $NaHCO_3$ solution and one tea spoonful activated charcoal. Shake the suspension for half an hour and then filter through a P free filter paper. Take 10 ml aliquot from the filtrate in a 25 ml volumetric flask, dilute to about 15 ml with DW and add a few drops of 2—4 DNP indicator which will turn the solution yellow. Add dilute HCl drop by drop to the solution till it turns colourless. Develop phospho-molybdic blue colour in 25 ml volume as stated earlier (sec. 2.2-13) and find out the optical density (OD) of the developed blue colour of the solution with the help of a colorimeter or spectrophotometer at 660 nm wavelength.

Calculation

If the colorimetric (OD) reading for the 25 ml volume with 10 ml aliquot corresponds to x ppm of P, then the concentration of Olsen's extractable (available) P in the soil will be $x \times 50$ ppm.

B. Total phosphorus

Various methods are available for extraction of total P from soils. More common are sodium carbonate (Na_2CO_3) fusion (Jackson, 1971) and HCl extraction (Piper, 1967). The second method being more simple to undertake is being discussed here.

Principle

Conc. HCI dissolves various minerals from the soil during the course of boiling which includes phosphorus as well. The concentration of P, thus extracted, can be determined colorimetrically through development of phosphomolybdic blue colour as stated earlier.

Requisites

(a) Apparatus : Hot plate, colorimeter/ spectrophotometer

(b) Glasswares : 500 ml conical flasks, 100 ml cylinder, 1 ℓ, 500ml, 250 ml, 25 ml vol. flasks, pipettes, funnels etc.

Reagents :
- (i) Conc. HCl
- (ii) 0.5 N HCl (approx) : Dilute 50 ml conc. HCl to 1 ℓ with DW.
- (iii) Reagents for P determination - as stated earlier.

Procedure

Take 20 g of soil sample in a 500 ml conical flask. Add 200 ml conc. HCl and mix thoroughly. Put a glass funnel on the mouth of the flask and boil for an hour on a hot plate. After digestion, cool the material and filter through a filter paper into another conical flask. Wash the residue a few times with hot 0.5 N HCl. Make up the volume of the extract to 500 ml in a volumetric flask and use it for determination of P.

Take 5 ml aliquot from tne extract and proceed for determination of P by following the method described for available P estimation.

Calculations

If concentration of P in 25 ml volumetric flask be x ppm, concentration of total P in soil will be x X 125 ppm.

C. Different inorganic forms

As stated earlier, P forms insoluble compounds by reacting with iron and aluminium under acidic condition and with calcium under alkaline environment. Anaerobic condition of the pond soil due to submergence tend to release considerable amount of P from such compounds through several mechanisms (Chattopadhyay and Mandal, 1980). Studies on inorganic forms of phosphorus may, therefore, be an important proposition for studying magnitude and dynamics of transformation of applied P in fish pond soils.

Principle

Chang and Jackson (1957) developed the methodology of fractionation of soil P into different inorganic forms using solubility of these compounds in various extractants. 1 (N) ammonium chloride (NH_4Cl) is used to extract saloid bound form of P which remains loosely held by soil particles while neutral N ammonium fluoride (NH_4F), 0.5 N H_2SO_4 and 0.1 N NaOH extracts Al, Fe and Ca bound forms of P respectively.

Requisites

(a) Apparatus: Colorimeter/spectrophotometer, mechanical shaker, centrifuge (preferably).

(b) Glasswares : Volumetric flasks 1 ℓ, 250 ml, 100 ml, 25 ml, narrow mouth 250 ml bottles, funnels, 150 ml conical flasks, pipettes etc.

(c) Reagents :

(i) NH_4Cl : Dissolve 53.5 g NH_4Cl in about 750 ml DW and make up the volume to 1 ℓ.

(ii) 0.5 N neutral NH_4F : Dissolve 18.5 g solid NH_4F in 1 ℓ. DW, adjust the pH to 7.0 with dilute NH_4OH and, if necessary, dilute HCl using a pH paper.

(iii) Saturated sodium chloride (NaCl): Put enough amount of NaCl to 1 ℓ DW in a wide mouth bottle so that some amount of the salt remains precipitated at the bottom. Do not remove the precipitate and use the solution as such.

(iv) 0.1 N NaOH : Dissolve 4g NaOH in 1 ℓ DW.

(v) 0.5 N H_2SO_4 : Dissolve 15 ml conc. H_2SO_4 into 750 ml DW, cool the solution, make up the volume to 1 ℓ

(vi) 0.8 M boric acid (H_3BO_3) : Dissolve 5.0 g boric acid in 1 ℓ DW.

(vii) Reagents for P estimation : As stated previously in sec. 2.2-13.

Procedure

Weigh 2 g soil sample and transfer to a 250 ml narrow mouth stoppered bottle. Add 100 ml of 1N NH_4Cl and shake for half an hour. Centrifuge the suspension for about 15 minutes at 2400 rpm and decant the supernatent solution. In case centrifuge is not available, keep the bottle horizontally for 6-7 hours which will allow the soil to precipitate down. Decant the supernatent solution and filter it through a filter paper keeping most of the soil samples retained in the bottle. Use the filtrate for estimation of saloid bound P by following standard procedure of P determination following sec. 2.2-13.

Put a funnel on the mouth of the bottle with saved soil sample and open the used filter paper on it. Take 100 ml of neutral N NH_4F in a cylinder, pipette out about 10 ml NH_4F from this cylinder and wash down the soil particles sticking on the filter paper to the bottle. Repeat this washing several times to wash the filter paper and then add the remaining portion of the NH_4F solution to the bottle. Shake the bottle for half an hour. Use the previous procedure to filter the solution containing Al bound form of P. Determine the P concentration in the solution by using standard method as stated earlier with the only exception that 8 ml 0.8 M boric acid should be added to the solution in the 25 ml volumetric flask before development of colour. Save the soil for determination of Fe bound form of P.

Wash the soils sticking with the filter paper into the bottle containing saved soil sample by using about 25 ml of NaCl solution. Shake the suspension, centrifuge or keep for a few hours in horizontal position and decant and discard the supernatent solution. Carry out this washing operation twice. Then shake the soil sample with 100 ml 0.5 N H_2SO_4 and proceed for filtration and subsequent washing of the soil with NaCl as stated earlier. Determine P content in the filtered solution to get the amount of Fe bound P.

Add 100 ml 0.1 N NaOH to the washed soil sample, shake for half an hour as stated earlier and filter the suspension. Determine P content of the filtrate to get the amount of P in Ca bound form.

Calculation

Find out the concentration of P in each extracted solution from the colorimetric reading by following sec. 2.2-13.

If the conc. of P in any solution be x ppm and the aliquot taken from the solution be y ml for making a total volume of 25 ml, amount of P in soil under each form of P will be $\frac{x}{y} \times 1250$ ppm

Comments

During determination of Al bound P, application of H_3BO_3 has been suggested. This is done to avoid the problem of interference of F^- ions with development of phosphomolybdic colour.

Levels of P in pond soils

Banerjea (1967) classified the available P status of fish pond soils of India with relation to their productivity. He suggested that available P_2O_5 below 30 ppm, 30-60 ppm, 60-120 ppm and above 120 ppm should be considered to be index of poor, average, good and high productivity.

In this book, determination of phosphorus has been expressed in terms of 'P'. To convert the value to P_2O_5, the result should be multiplied by 2.29.

3.2-12 Potassium (Available)

In spite of being the third primary nutrient element influencing productivity of primary fish food organisms, importance of potassium (K) has not yet received much attention. In general, fish ponds exhibit good occurrence of K and availability of this nutrient element does not usually create much problem. However, in case of ponds having bottom soils with poor K content or where high doses of other nutrient elements have been added, availability of K may appear to be an important factor (Dobbins and Boyd, 1976; Chattopadhyay, 1995).

Principle

Readily available K for growth and production of primary fish food organisms occurs mainly in two forms in pond soils viz. K in soil solution and K in exchangeable form. Estimation of available K in pond soil, therefore, should include both of these two forms.

In sec 3.2–8 determination of exchangeable K has been described. Same principle may be adopted for determining available K status of pond soils with the only exception that water soluble portion of K is not removed during the determination.

Requisites

(a) Apparatus : Mechanical shaker, flame photometer

(b) Glasswares : 1 *l* vol. flask, 100 ml cylinder, 100 ml bottles.

(c) Reagents : N neutral ammonium acetate (CH_3COONH_4) - Prepare the reagent by following sec. 3. 2- 8.

Procedure

Take 5 g soil in a 100 ml bottle, add 50 ml neutral N CH_3COONH_4, stopper the bottle and shake for half an hour. Filter the suspension and determine K content of the filtrate with the help of a flame photometer by following sec. 3.2–8.

Calculations

If concentrations of K in the filtrate is x ppm, amount of available K in the soil will be $x \times 10$ ppm.

Comments

If watersoluble and exchangeable forms of K are to be determined separately, 10 g soil sample may be shaken with 20 ml DW and the filtrate be used for determination of watersoluble K.

Amount of exhangeable K can be estimated by following sec. 3.2.8 or by subtracting the value of watersoluble K from that of water-soluble + exchangeable form.

3.2-13 Soil Texture

The objective of determining textural composition of soils is to know the percentage of soil materials contained in different grain size fractions viz. sand, silt and clay and to classify the soils to any particular textural group so that the dominant grain sizes present in the soil can be identified easily.

Principle

Textural composition of soils can be estimated by means of mechanical analysis. This consists essentially of two distinct operations viz. dispersion of the soil to ultimate grain sizes and grading the dispersed grains according to their groups.

Requisites

(a) Apparatus : Mechanical shaker, stirrer.

(b) Glasswares : 1 *l* cylinder, 20 ml pipette, 500 ml and 50 ml beaker, 250 ml conical flask etc.

(c) Reagents :

(i) 6 per cent hydrogen peroxide (H_2O_2) : H_2O_2 is generally sold at 30 per cent concentrations. For the purpose of preparing 6 per cent H_2O_2, this 30 per cent solution is to be diluted at 1 : 5 H_2O_2 : Water ratio.

(ii) 2 N HCI : Dilute 100 ml concentrated HCl to 500 ml for getting approximately 2 N HCl.

(iii) 2 N NaOH : Dissolve 40 g NaOH pellets in about 300 ml DW. Allow it to cool and dilute upto 500 ml with water.

(iv) 5 per cent silver nitrate ($AgNO_3$) : Dissolve 5 g $AgNO_3$ in 100 ml DW.

Procedure

Take 20 g of soil in a 50 ml beaker and add about 250 ml DW. Boil it for about 10-15 minutes and decant the supernatent water.

Now add 30 ml of 6 per cent H_2O_2 to this soil sample and digest on a water bath. Add more H_2O_2 if necessary and continue digestion till the frothing stops. Add about 30 ml 2 N HCl and 100ml DW, boil for a few minutes and allow the soil suspension to stand for one hour with frequent stirring to make the soil free from carbonates. Filter the soil and wash it repeatedly wih hot water till the filtrate becomes chloride (Cl^-) free. This can be tested by receiving a few drops of filtrate in a test tube containing small amount of $AgNO_3$. Presence of Cl^- will develop white colours in the solution. Transfer the suspension to a 250 ml conical flask, add 5 ml of 2 N NaOH and shake for half an hour with a mechanical shaker. Transfer the contents immediately to a 1000 ml tall cylinder, make up the volume with DW and stirr the entire volume thoroughly with the help of a stirrer. Allow the suspension to stand for exactly 4 minutes, immediately lower a 20 ml pipette to 10 cm depth of the suspension and collect 20 ml of the content. Transfer it in a 50 ml previously weighed beaker marked as beaker *A*. Make up the volume of the 1000 ml cylinder with DW, stirr the suspension as before and allow it to stand for 6 hours. Repeat the same procedure to collect 20 ml of suspension to a previously weighed beaker marked as *B*. Dry the suspension in two beakers in an oven to constant weights. Get the actual weights of dry materials in the beakers after subtracting the weights of the empty beakers from the values of beakers + dried materials.

Calculation

If the weight of dry material in beaker A be x g and that in beaker B be y g

then

Percentage of clay = $y \times 250$

Percentage of silt = $(x - y) \times 250$

Percentage of sand = $100 - (x \times 250)$

Comments

The values of sand, silt and clay obtained through mechanical analysis of soil sample can be utilised to know the textural classification of the soils. This can be achieved by using per cent values of sand, silt and clay with figure - 3 and finding the respective textural class.

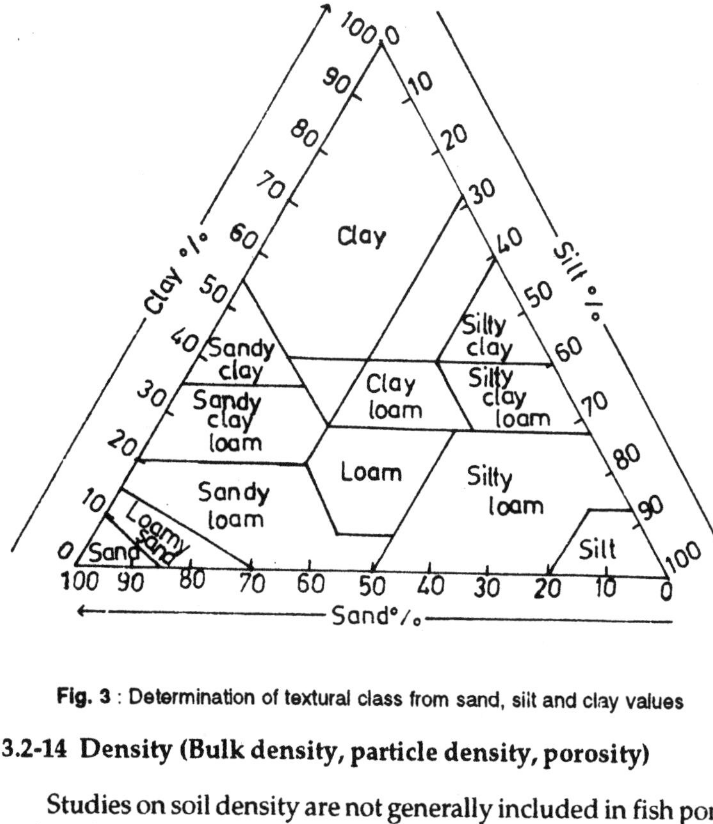

Fig. 3 : Determination of textural class from sand, silt and clay values

3.2-14 Density (Bulk density, particle density, porosity)

Studies on soil density are not generally included in fish pond soil analysis programmes. However, these studies have special significance for understanding some physical qualities of pond soils viz. occurrence of pore space, movement of water etc. Bulk density refers to the mass of water free soil to its bulk volume while particle density is the average density of soil particles (USDA, 1968). Bulk density is also called apparent density and particle density is sometimes referred to as real density of soils.

Principle

Bulk density (BD) of a soil sample can be estimated by

determining the dry weight of the soil and dividing it with the volume of field moist soil. For determination of particle density (PD), on the other hand, difference between the weight of a container when completely filled with water and when filled partially with a weighed soil sample and the rest with water is used.

Requisites

(a) Apparatus : Metalic core sampler of known volume (around 50 ml), oven.

(b) Glasswares : 100 ml bottle.

Procedure

A. Determination of BD

Weigh the core sampler (W_1g). Drive the sampler into the soil surface and take it back. Wipe the mouth of the sampler to collect soil sample having same volume as the core sample (V ml). Dry the soil sample inside the core sampler in oven at around 105°C till a constant weight is obtained. Weigh the oven dry soil alongwith sampler (W_2g).

B. Determination of PD

Weigh a 100 ml bottle under empty condition (M_1g) and also when completely filled with water (M_2g). Drain the water and dry the bottle. Fill it partially with some dry soil sample and get the weight of the bottle when partially filled with soil (M_3g). Pour water to the remaining portion of the bottle so that the bottle becomes completely filled with soil + water (M_4g).

Calculations

$$BD = \frac{W_2 - W_1}{V} \text{ g cc}^{-1}$$

$$PD = \frac{(M_3 - M_1)}{(M_2 + M_3 - M_1 - M_4)} \text{ g cc}^{-1}$$

Porosity

Porosity is the amount of soil space not occupied by soil particles and is usually expressed in terms of percentage porosity. This can be calculated from BD and PD values of soils using the following formulae:

$$\text{Porosity (\%)} = \frac{PD - BD}{PD} \times 100$$

3.2-15 Water holding capacity

Water holding capacity (WHC) of a soil denotes the amount of water that can be retained by unit volume of soil. This parameter should be attended carefully when a new site is brought under fish cultural activities or for the ponds having problem of water retention.

Principle

For determining WHC of a soil, a weighed volume of dry soil is incubated with water for sufficient time so that the water is absorbed freely by the soil. After the lapse of a certain period, the total weight of the soil alongwith absorbed moisture is obtained and the WHC of the soil is calculated therefrom.

Requisites

(a) Apparatus : Circular metalic boxes (Keen's box)- These are simple circular aluminium boxes with lids, perforated at the bottom. Generally these boxes have 2.8 cm internal radius and 1.6 cm height to provide an internal volume of about 40 ml (39.42 ml)

(b) Glasswares : Petridish.

Procedure

Place a filter paper, cut to the size of the base area of the Keen's box, to the bottom of the box. Weigh the box alongwith the filter paper (W_1g). Add soil samples to the box in small instalments with continuous tapping so that the soil occupies the volume of the box uniformly. Fill the box in the same manner and level the upper surface with the help of a spatula or knife.

Keep the box on a petridish and fill the petridish partially with water so that about $^1/_3$ rd of the height of the box remains dipped in water. Leave the box for overnight. Then take it out, wipe the outside surface to remove any stain of water and weigh it again (W_2g). Place the box in an oven at about 105°C and dry it till a constant weight is obtained (W_3g).

Take five nos of filter papers of same quality used in this study and cut to the size of the base of the box. Weigh them together, dip into water once, let the excess water drop down and weigh immediately the moist filter papers. Divide the obtained weight by five to get the weight of one moist filter paper. Deduct the dry weight of individual filter paper and find out the amount of moisture absorbed by a single filter paper (W_4g).

Calculations

If

Wt of the box + filter paper is W_1g

Wt of the box + filter paper + dry soil is W_2g

Wt of the box + filter paper + moist soil is W_3g

and water absorbed by single filter paper is W_4g

Then

Wt of dry soil = W_2 - W_1 g

Wt of water absorbed by the soil = W_3-W_2-W_4 g

and

$$\text{WHC of the soil} = \frac{W_3 - W_2 - W_4}{W_2 - W_1} \times 100\%$$

APPENDICES

APPENDIX–I

Preparation of Some Standard Solutions

A. Standard Solutions

A standard solution is one whose exact strength is known. Some of these solutions contain known concentrations of compounds having definite composition and hence their strengths need not be checked or standardised after preparation and are known as primary standards. For example, 53 g of dried sodium carbonate (Na_2CO_3) dissolved in 1 ℓ DW will make a Na_2CO_3 solution of 1 N strength. However, there are some other substances which are not always obtained in exact composition owing to several reasons like instability, variability and solutions of such substances need to be standardised with another standard solution. These are known as secondary standard solutions. Sulfuric acid (H_2SO_4) forms an example of such solution whose strength is ascertained with the help of a standard base solution, after preparation of a solution of desired approximate strength.

B. Preparation of Primary Standard Solutions

1 N Na_2CO_3 solution : Equivalent weight of Na_2CO_3 is 53 g. Hence weigh accurately 53 g of analytical grade, dried Na_2CO_3 and wash down to a 1ℓ volumetric flask. Add DW to dissolve the Na_2CO_3 and make up the volume to 1 ℓ. This will give a standard 1 (N) solution of Na_2CO_3. This is a primary standard solution and need not be standardised again.

0.1 N Na_2CO_3 : For preparation of a 0.1N or N/10 Na_2CO_3 solution, either dissolve 5.3 g Na_2CO_3 in 1 ℓ DW as before or dilute the previously prepared 1 N solution 10 times.

(0.02)$_2$ N Na_2CO_3: Similarly, for preparing 0.02N or N/50 Na_2CO_3 solution, 0.1 N Na_2CO_3 solution should be diluted 5 times.

C. Preparation Secondary Standard Solution

1 N H_2SO_4 Solution : H_2SO_4 produces secondary solution and hence it should be standardised with a standard base. Conc. H_2SO_4 usually has a strength of 36 N. Use this concentration to prepare a H_2SO_4 solution whose approximate strength will be slightly higher than the desired one. Standardise this solution with an 1 N standard base and dilute accordingly to get 1N solution of H_2SO_4.

For any standardisation, use the equation

$$V_1S_1 = V_2S_2$$

where

V_1	=	Volume of the standard solution
V_2	=	Volume of the unknown solution
S_1	=	Normality of the standard solution
S_2	=	Normality of the unknown solution.

Concentration of H_2SO_4 usually has an approximate strength of 36 N. Use this concentration to prepare a H_2SO_4 solution whose approximate strength will be little more than 1N.

e. g. $V_1S_1 = V_2S_2$

If 35 ml conc. H_2SO_4 (V_1) of 36 N strength (S_1) is diluted to 1000 ml (V_2), then the strength of the diluted solution (S_2) will be

35 x 36/1000 = 1.26 N.

However, since the strength of conc. H_2SO_4 is not exactly known, this normality of the solution will be only approximate and to get the actual strength of this diluted H_2SO_4, it should be standardised against a standard base.

Take 10 ml (V_2) of the H_2SO_4 solution (S_2) in a 250 ml conical flask. Put a few drops of methyl orange indicator, add a little DW and titrate with 1N Na_2CO_3 (S_1) to a faint yellow colour. If the volume of standard (1N) Na_2CO_3 solution required for titration is V_1ml, then the strength of the acid may be calculated to be as follows :

Here, V_1 = titre value (say 11 ml in this titration)

S_1 = 1 N

V_2 = 10 ml (Vol. of H_2SO_4 used)

S_2 = unknown.

Hence

$$S_2 = \frac{V_1S_1}{V_2} \text{ or } S_2 = \frac{11\times 1}{10} = 1.1$$

i.e. strength of the diluted H_2SO_4 is 1.1 N.

Now, to prepare an 1 N solution of the acid, use the same equation again.

V_1 = Vol. of 1 N H_2SO_4 required (say, 1000 ml)

S_1 = Required strength of H_2SO_4 (1 N here)

V_2 = Vol. of the 1.1 N H_2SO_4 to be dilute (unknown)

S_2 = Strength of this H_2SO_4 (1.1 N)

Here

$$V_2 = \frac{V_1S_1}{S_2} \text{ or } \frac{1000\times 1}{1.1} = 909.1 \text{ ml.}$$

Hence to get an 1 N strength, 909.1 ml of the prepared 1.1 N solution of H_2SO_4 should be diluted to 1000 ml.

0.1 N H_2SO_4 : 0.1 or N/10 solution of H_2SO_4 can be prepared by diluting the standard 1N solution 10 times. This can be achieved by preparing a dilute H_2SO_4 solution which will have a strength of slightly more than 0.1 N and then proceeding as stated earlier.

0.02 N H_2SO_4 : 0.02 N or N/50 H_2SO_4 can be prepared by diluting 0.1 N H_2SO_4 5 times with DW or from a straightly prepared solution of approximate (slightly high) 0.02 N H_2SO_4, as discussed previously.

However, in each case, these solutions should be standardised by titrating with any standard alkali solution.

Standard NaOH

For preparing standard NaOH solutions the same method should be followed using equivalent weight of NaOH.

Strengths of some common acid and bases have been given in Appendix - II and general guidelines or preparing dilute solutions have been given in Appendix–III.

APPENDIX—II

Strengths of some common acids and bases under concentrated forms.

	Acid/base	*Normality*	*%*
1.	Hydrochloric Acid (HCl)	11.6	37 — 38
2.	Acetic Acid (CH_3COOH)	17.5	99.5
3.	Nitric Acid (HNO_3)	16.0	70 —71
4.	Sulphuric Acid (H_2SO_4)	36	97 — 100
5.	Perchloric Acid ($HClO_4$)	9—11.6	60 — 70
6.	Orthophosphoric Acid (H_3PO_4)	45	85
7.	Ammonium Hydroxide (NH_4OH)	15	28

APPENDIX—III

Preparation of dil. solution from conc. ones.

	To get approximately 1 N solution of	*Dilution procedure*
1.	Hydrochloric Acid (HCl)	Dilute about 86 ml. conc. HCl to 1 ℓ with DW.
2.	Acetic Acid (CH_3COOH)	Dilute about 57 ml CH_3COOH to 1 ℓ with DW.
3.	Nitric Acid (HNO_3)	Dilute about 62 ml HNO_3 to 1 ℓ with DW.
4.	Sulphuric Acid (H_2SO_4)	Dilute 28 ml H_2SO_4 slowly and cautiously to about 450 ml DW with constant stirring, cool the solution and make the volume to 1 ℓ.
5.	Ammonium Hydroxide (NH_4 OH)	Dilute about 57 ml NH_4OH with DW to 1 ℓ
6.	Sodium Hydroxide (NaOH)	Dissolve 40 g NaOH pellets/flakes in about 750 ml DW, cool and dilute to 1 ℓ.

APPENDIX—IV

Some important measurements.

1 ha	=	2.5 acre
1 acre	=	0.40 ha
1 ha	=	10,000 m^2
1 ft^2	=	0.093 m^2
1 m^2	=	10,000 cm^2
1 m^2	=	10.76 ft^2
1m^3	=	1000 ℓ
1 ham	=	10,000 m^3
1 acre ft	=	1233.6 m^3

REFERENCES

Alikunhi, K.H. (1957). Fish Culture in India. Farm Bulletin (20). ICAR. New Delhi pp 144.

Anon. (1982). Annual Report. CIFRI, Barrackpore.

A.P.H.A. (1971). Standard Methods for the Examination of Water and Wastewater. Washington D.C pp 874.

Banerjea, S. M. (1967) water quality and soil condition of fish ponds in some states of India in relation to fish production. Indian J. Fish. 14 : 115–144.

Banerjea, S.M. and Mandal, L.N. (1965). Inorganic transformation of watersoluble phosphate added to fish ponds as influenced by the nature of pond soils. J. Indian Soc. Soil Sci. 13 : 167-137.

Banerjee, R.K. and Chattopadhyay, G.N. (1980). Methodology of Soil and Water Analysis in Brackishwater Aquaculture System. Misc contribution no. 15, CIFRI, Barrackpore, India.

Black, C.A. (1965) Methods of Soil Analysis. American Soc. Agron. Madison.Wisconsin. USA.

Boyd, C.E. (1974). Lime Requirements of Alabama Fish Ponds. Auburn Univ. (Ala) Agri. Expt. Stn. Bull. (459). pp 20.

Boyd, C.E. (1978). Water Quality in Warmwater Fish Ponds. Agril. Expt. Stn. Auburn Univ. pp 359.

Boyd,C.E. (1979). Aluminium sulphate (alum) for precipitating clay turbidity from fish ponds. Trans. Amer. Fish Soc. *108* : 307–313.

Boyd, C.E. (1982). Water Quality Management for Pond Fish Culture. Elsevier Scientific Publishing Co. New York pp 318.

Buttery, B.R. and Buzell, R.I. (1977). On relationship between chlorophyll content and rate of photosynthesis in soyabeans. Cam. J. Pl. Sci. *57* : 1-5.

Chang, S.C. and Jackson, M.L. (1957). Fractionation of soil phosphorus. Soil Sci. *64* : 133-144.

Chattopadhyay, G.N. (1995). Use of potassium in aquaculture. In: G.N.Chattopadhyay (Ed.) Nutrient Management in Aquaculture. Soil Testing Laboratory. Institute of Agriculture. Visva-Bharati. Sriniketen, India. pp 38-41.

Chattopadhyay, G.N. and Mandal, L.N. (1980). Inorganic transformation of applied phosphorus in brackishwater fish pond soils of West Bengal. Hydrobiologia *71* (1-2) : 125—130.

Chattopadhyay, G.N., Roy, A.K., Das, P.R. and Banerjee, R.K.(1985). A note on the determination of available nitrogen in brackishwater fish pond soils. J. Indian Soc. Coastal Agric. Res. *3* (1) : 673- 66.

Chattopadhyay, G.N., Roy, A.K. and Das, P.R. (1991). A note on determination of available phosphorus in brackishwater fish pond soils. J. Indian Soc. Coastal Agric. Res. *9* (1-2) : 77-79.

Chattopadhyay, G.N.; Saha, P.K.; Ghosh A. and Karmakar, H.C. (1988) A study on optimum BOD levels for fish culture in wastewater ponds. Biol. Wastes. 25 : 79-83.

CIFRI (1985). Carp Culture— Package of Practices for Increasing Production. Aquaculture Extn. Manual (new series no. 2). CIFRI, Barrackpore, India.

Dev, G (1995). Fertilisers and their requirements in fish ponds. In : G.N. Chattopadhyay (Ed). Nutrient Management in Aquaculture. Soil Testing Laboratory. Institute of Agriculture, Visva-Bharati. Sriniketan. India. pp 54-62.

Dickman, S.R. and Bray, R.H. (1940). Colorimetric determination of phosphate. Ind. Eng. Chem. Anal. *12* : 665-668.

Dobbins, D.A. and Boyd, C.E. (1976). Phosphorus and potassium fertilisation of sunfish ponds. Prog. Fish Cult. 27 : 75-78.

Dust, J.V. and Shindala, A. (1970). Relationship of chlorophyll-a to algal count and classification in oxidation ponds. J. Water Pollution Control Fed. *42* : 1362-1369.

Ellis, M.M. (1937). Detection and Measurement of Stream Pollution. Bull 22, US Bureau of Fisheries.

Ghildyal, B.P. (1976). Fertility and properties of submerged rice soils. In : J.S. Kanwar (Ed.) Soil Fertility, Theory and Practice, ICAR. New Delhi.

Ghosh, A.N.; Ghosh, S.R. and Sarkar, N.N. (1973). On the salinity tolerence of fry and fingerlings of Indian major carps. J. Inland Fish. Soc. India V : 215-217.

Golterman, H.L. (1967). Influence of the soil on the chemistry of water in relation to productivity. F.A.O. Fish Rep. (3) : 27-42.

Golterman, H.L.; Clymo, R.S. and Ohnstad M.A.M. (1978). Method for Physical and Chemical Analysis of Fresh Waters. Blackwell Scientific Publications London. pp. 211.

Hickling, C.F. (1971). Fish Culture. Faber and Faber. London pp 225.

Irwin, W.H. (1945). Methods of Precipitating Colloidal Soil Particles from Impounded Waters of Central Oklahoma. Okla. Agr. Mech. Coll. Bull no. 42. pp. 16.

Jackson, M.L. (1971). Soil Chemical Analysis. Prentice-Hall of India Pvt. Ltd. New Delhi. pp. 498.

Jhingran, V.G. (1975). Fish and Fisheries of India. Hindustan Publ. Corp. New Delhi pp. 954.

Jhingran, V.G.; Natarajan, A. V.; Banerjea, S. M. and David, A. (1968). Methodology on Reservoir Fisheries Investigations in India. Bull (12). CIFRI, Barrackpore, India. pp 102.

Mandal, L.N. and Chattopadhyay, G.N. (1990). Chemical and electro-chemical environment in fish ponds and it's impact on aquaculture. In : G.K. Manna and B.B. Jana (Eds.). Impact of environment on animals and aquaculture. Univ. of Kalyani, W.B., India, pp 79-84.

Mandal, L.N. and Chattopadhyay, G.N. (1992). Nutrient management in aquaculture, In : H.L.S. Tandon (Ed.) Non-Traditional Sectors for Fertiliser Use. FDCO. New Delhi, India pp 1-17.

McKee, J.E. and Wolf, H.W. (1963). Water Quality Criteria. State of Calif. State Water Quality Control Board. Publ. no. 3—A Sacramento. pp 548.

Murphy, J. and Riley, J.P. (1962). A modified single solution method for the determination of phosphate in natural waters. Analytica Chim. Acta. 27 : 31.

Neess, J.C. (1949). Development and status of pond fertilisation in Central Europe. Trans. Amer. Fish. Soc. *76* : 335-358.

Odum, E.P. (1971). Fundamentals of Ecology. Saunders College Publishing. Philadelphia. pp 574.

Olsen, S.R.; Cole, C.V.; Watanabe, F.S and Dean, L.A. (1954). Estimation of available phosphorus in soils by extraction with sodium bicarbonate. US Dept. Agric. Circ. 939.

Parks, R.W.; Scarsbrook, E. and Boyd, C.E. (1975) Phytoplankton and Water Quality in a Fertilised Fish Pond. Auburn Univ. (Ala.) Agr. Exp. Sta., Cir. 224. pp 16.

Pearsall, W.H. and Mortimer C.H. (1939). Oxidation-reduction potentials in waterlogged soils, natural waters and muds. J. Ecol. 27 : 483-501.

Piper, C.S. (1967). Soil and Plant Analysis. Hans Publishers, Bombay.

Santhanam, R.; Velayutham, P.and Jegatheesan, G. (1989). A Manual for Freshwater Ecology, Daya Publishing House, Delhi pp. 134.

Sawyer, C.N. and McCarty, P.L. (1967). Chemistry for Sanitary Engineers. Mc Graw-Hill Book Co. New York, pp 518.

Schaeperclaus, W. (1933). Text Book of Fish Centure. U.S. Fish and Wild Serv. pp 26.

Singh, S. and Ram, H. (1971). A comparative study on pond and adjoining cultivated soils. Indian J. Agric. Chem. 4 (1) : 13-20.

Sreenivasan, A. (1968) The limnology of and fish production in two ponds of Chinglepet (Madras). Hydrobiologia 32 : 131-144.

Subbiah, B. and Asija, G. L. (1956). A rapid procedure for estimation of available nitrogen in soils. Curr. Sci. 25 (8) : 259.

Swingle, H.S. (1947). Experiments on Pond Fertilization. Alabama Polytech. Institute Agril. Expt. Stn. Bull. no. 264. Auburn, USA. pp 36.

Theroux, F.R.; Eldridge, E.F. and Mallman, W, LeRoy (1943). Laboratory Manual for Chemical and Bacterial Analysis of Water and Sewage. McGraw Hill Book Co. Inc. London, pp 274.

USDA (1968). Diagnosis and Improvement of Saline and Alkaline Soils. Agril. Handbook no. 60, USDA, Oxford and IBH Publ. Co. Calcutta. pp 156.

Vollenweider,. R.A. (1969). A Manual on Methods for Measuring Primary Production in Aquatic Environments. IBP Hand Book No. 12. Blackwell Sci. Pub. pp. 213.

Walkley, A. and Black, I. A. (1934). An examination of the Degrjareff methd for determining soil organic matter, and a proposed modification of the chromic acid titration mehod. Soil Sci. *37* : 29-38.

Wallen, I.E, (1951). The Direct Effect of Turbidity on Fishes. Okla. Agr. Mech. Coll. Bull 48 : pp 27.

White, C.E. Jr. (1955) Chlorine : It's Toxicity to Gold fish, Fathead Minnows, Golden Shiners and Bluegills and It's Removal from water. Ms thesis. Albama Poly Inst. Auburn pp 58.

Winkler, L.W. (1888). The Determination of dissolved oxygen in Water. Berline Deutsch Chem Gesellsch *21* : 2843.

Wolny, P. and Grygierek (1972). Intensification of fish ponds production. In Z.Kajak and A. Hillbricht Il kowska (Eds). Productivity Problems of Freshwater. Polish Scientific Publishers. Warshaw. pp. 563-571.

Yadav, B.R. and Khera, M.S. (1993) Analysis of irrigation waters In: HLS Tandon (Ed.) Methods of Analysis of Soils, Plants, Waters and Fertilisers. FDCO. New Delhi pp 144.

INDEX